FROM DIVINE COSMOS TO SOVEREIGN STATE

FROM DIVINE COSMOS TO SOVEREIGN STATE

An Intellectual History of Consciousness and the Idea of Order in Renaissance England

Stephen L. Collins

OXFORD UNIVERSITY PRESS
New York Oxford

Oxford University Press

Oxford New York Toronto
Delhi Bombay Calutta Madras Karachi
Petaling Jaya Singapore Hong Kong Tokyo
Nairobi Dar es Salaam Cape Town
Melbourne Auckland

and associated companies in
Berlin Ibadan

First published in 1989 by Oxford University Press, Inc.,
200 Madison Avenue, New York, New York 10016
First issued as an Oxford University Press paperback, 1991
Oxford is a registered trademark of Oxford University Press

Library of Congress Cataloging-in-Publication Data
Collins, Stephen L.
From divine cosmos to sovereign state:
an intellectual history of consciousness and
the idea of order in Renaissance England /
Stephen L. Collins.
p. cm. Bibliography: p. Includes index.

1. Great Britain—Intellectual life—16th century 2. Great Britain—Intellectual life—17th century.
3. Great Britain—Politics and government—1485-1603.
4. Great Britain—Politics and government—1603-1647.
5. Philosophy, English—16th century. 6. Philosophy, English—17th century.
7. Cosmology—History. 8. Consciousness—History.
9. Order—History. I. Title.
DA320.C65 1989 941.05—dc 19 88-4225
ISBN 0-19-505458-X
ISBN 0-19-507131-X (pbk)

1 3 5 7 9 8 6 4 2

Printed in the United States of America
on acid-free paper

To Blanche and David, my parents,
for their sustaining love
and to
Morris

PREFACE

My interest in the intellectual history of the English Renaissance and in Shakespeare in particular began during a summer matinee performance at the Aldwych Theater, London, 1969. My experience of *Troilus and Cressida* that afternoon exhilarated yet disconcerted me. While I had never formally studied Shakespeare, I sensed that I had experienced him "wrongly"—not as I was supposed to, not as "intended." As I turned to Elizabethan historians and literary critics and returned to *Troilus and Cressida* (at the theater *and* in the text), I began to feel that the unease I had experienced, the dis-order (if you will), was meaningful.

The play I watched and listened to and the text I read were problematical. Its language gave the lie to its "meaning", to the "ideology" of order that Tillyard[1] and others so clearly—and so appropriately, it appeared—identified as a thematic Elizabethan and Shakespearean convention. It was not that the play (and as I continued to read other of Shakespeare's plays, I saw a similar relationship) was not about order, it was more that it problematically questioned its own "obvious" theme. It simultaneously established (reinforced, perhaps) and contested (critiqued) an "ideology" of order. Such textual self-contestation denied reductive interpretations that identified objective meaning in Shakespearean text and/or in Elizabethan cultural contexts; it suggested that historical meaning could best be apprehended as an experiential process.

In trying over the years and in this book to approach the historical nature of this textually momentous self-contestation, I have treated the idea of order both as the objectively conventional Elizabethan ideology and as the referential context against which such "meaning" is subjectively problematized. As an intellectual history of the idea of order, this book hopes in the words of Dominick LaCapra "to formulate as a problem what is so often taken deceptively as a solution: the relationship between texts and their various pertinent contexts."[2]

To do this, the problematical nature of the idea of order must be respected and simple interpretive reversal avoided. Recent scholars (especially literary critics) have busily refuted an Elizabethan or a Shakespearean idea of order and have highlighted the marginal, often at the utter expense of and out of context from the conventional.[3] To isolate the anomalous or the conventional is ahistorical as well as reductive. As LaCapra suggests, we should emphasize what is marginal "as seen from the conventional view,"[4] but we must do this in a way that neither cancels conventional reality nor raises the marginal to an absolute level.

I have tried to contextualize (conceptualize) the idea of order problematically (historically) by examining the relationship between the marginal (anomalous) and the conventional (paradigmatic). In so doing I indicate how the marginal becomes first more—then less—self-conscious and indeed becomes "meaningful" (comes to terms with the conventional), while the conventional also becomes more self-conscious and less taken for granted.[5] Highlighting the marginal, then, is not to isolate it, or to emphasize it nonproblematically, or to claim its conventionality. Rather, it is to interpret the historical process of its becoming meaningful—to experience the idea of order as a problem, not as a solution.

I have treated the idea of order as a dynamic concept, one that was changing even as it articulated stasis. And I have iterated and reiterated the becoming self-consciousness of attitudes that characterized this dynamic quality. I am not suggesting that these attitudes never existed prior to the late sixteenth century—nor, of course, am I claiming that the traditional idea of order described a uniformly coercive and coerced social reality. I am contextualizing the historical process of meaning redefinition—the making programmatic (author-izing) that which existed unself-consciously as a critique of the taken-for-granted. In this same way the relation between the marginal and the conventional indicates the existentially reflective nature of modernity, which "becomes" most self-consciously articulated at the end of the nineteenth century just as the traditional Tudor idea of order became self-consciously elaborated as it was no longer taken for granted.

I would like to thank Babson College's Board of Research for funds that helped with the expense of preparing the manuscript, and Joan Walter, Victoria Delbono, Concetta Stumpf, and Barbara Raymond who graciously and efficiently word processed it over a period of two years—not always under the easiest circumstances. Thanks as well are due to the library staffs at the British Museum Reading Room, Wellesley College, and Babson College for their help. Over the years colleagues at Shimer College, Boston University, and especially at Babson College, have shared ideas and supported my endeavors. I thank them.

Most important, however, has been Diane's confidence in me, without which I might have chosen an easier path. With love, I thank her.

I have maintained original spelling wherever possible; the English Renaissance sounds better that way.

Wellesley, Mass. S.L.C.
March 1988

CONTENTS

Introduction. Thematic, Methodological, and Organizational Considerations, 3

1. From Divine Cosmos to Sovereign State, 14
   The Tudor Idea of Order—fully described.
   The Idea of Order in the 1640s—outlined.

2. The Dramatic Microcosm: From Tamburlaine to Prospero, The Struggle with Self-Definition, 40
   Tudor Drama and Marlowe.
   Shakespeare.

3. Anomalies and Alternatives in Elizabethan England, 71
   The Subtle Changes in the Order Men: Barckley, Merbury, Mulcaster, Smith, Wentworth, et al.
   Richard Hooker: The Innovative Conservative.
   The Doleman Controversy: Catholic Political Psychology.

4. Refining and Defining: Jacobean England, 109
   The New Order Theorists and Radical Conservatism: James I, Forset.
   The Secular Inroads, Political Psychology, and New Values in Politics and Economics: Raleigh, Bacon, Eliot, and Merchant Pamphlets.
   The Hermetic-Eirenicist View: Sidney, Greville, and the Cult of the Magus. The Place of Nature, Philosophy, and Magic in New Social Utility: Sir Francis Bacon.

5. The Final Defining Voices, 149
   Sir Robert Filmer and John Selden.
   Pamphlets on Politics and Society, 1640–1643: Parker, Herle, Hunton vs. Digges and Ferne.
   Hobbes and the Sovereign State.

Conclusion. The Self-Consciousness of Consciousness: The Perceived Order of Modernity, 165

Notes, 169

Bibliography of Works Cited, 215

Index, 225

FROM DIVINE COSMOS TO SOVEREIGN STATE

INTRODUCTION

Thematic, Methodological, and Organizational Considerations

Perhaps there is no better dramatic expression of Tudor commonplaces and of the Tudor idea of order than Norton and Sackville's *Gorboduc.* Written in 1561–1562, in the dawning years of "Astraea,"[1] *Gorboduc* is about unity, about order, about degree. It is not a tragedy about person or about transcendence but a tragedy about social disorder and civil war. Implicit in the very affirmation of commonplace attitudes and concepts is an incipient insecurity about the viability of these commonplaces. Yet the authors reveal no need to question the basic characteristic relations in society. There is no self-conscious anxiety, no restlessness about purpose or meaning in life. While the play is concerned with preserving the life of the "body politic," it is never concerned with personal mortality.[2] Redemption seems secure. Indeed, the promise of divine retribution with which *Gorboduc* ends expresses in an important way the basic understanding of man and of society that defined the commonplace Tudor idea of order:

> But now, O happy man, whom speedy death
> Deprives of life, ne is enforc'd to see
> These hugy mischiefs, and these miseries,
> These civil wars, these murders, and these wrongs,
> Of justice, yet must God in fine restore
> This noble crown unto the lawful heir:
> For right will always live, and rise at length,
> But wrong can never take deep root to last.[3]

The same cannot be said about Christopher Marlowe's plays, written twenty-five to thirty years later. The question that *Doctor Faustus* raises is this: if the end of life is death, are life and its goals meaningless? Faustus unhappily contemplates mortality. When he is weighing his professions, medical and theological, he bemoans:

> Yet art thou still but Faustus and a man.
> Couldst thou make men to live eternally,
> Or, being dead, raise them to life again,
> Then this profession were to be esteemed . . .

> Why then belike we must sin,
> And so consequently die.
> Ay, we must die an everlasting death . . .
> What will be, shall be! Divinity, adieu![4]

Certainly the redemptive institutions of society are less operative than they had been. The energy and volition with which Marlowe's characters seek meaning for their lives are expressive of a spontaneous reaction to the awareness that the terms necessary for personal salvation need redefinition. Each main character chooses his own medium for salvation, but defines it so absolutely and submits to it so fully, that instead of attaining the desired ends, he creates new secular limitations and, in a larger sense, a new system of damnation. Faustus, Tamburlaine, and Barabas, each creates his own nemesis. There is no Christian retributive force at work in Marlowe's theatrical world, as there was in Norton and Sackville's. Finally, Marlowe's characters fail to secure salvation, from either traditional sources or their own endeavors. Indeed they lose sight, in midjourney, of their aspired-for destinations. Yet they do begin to define the boundaries of a new, modern arena in which individuals attempt to locate meaning in a secular and temporal world.

Marlowe is certainly not alone in contemplating mortality. Shakespeare, Donne, and Raleigh are but three writers conspicuously concerned with death and time. Yet, perhaps, only Thomas Hobbes attempted to base a political psychology of order upon a concern with mortality. In the traditional Tudor world, personal and social order were geared toward the *summum bonum*. For Hobbes, it has been argued, the motivating force behind social order is the fear of death. Order, rather than being geared toward the *summum bonum*, is defined by a fear of the *summum malum*.[5] Hobbes' preoccupation with order can, in this light, be seen as a profound concern to secure the institutional forces of redemption.

An intellectual history of the relationship between consciousness and order in late sixteenth- and early seventeenth-century England details the substitution of the secular sovereign state for traditional redemptive institutions. As old forces for redemption function less adequately, social reality appears less cohesive and more coercive. Mortality becomes problematical. This results in personal anxiety and social anomie and is characterized by a cultural process of meaning redefinition. In this context "meaning" must be understood as a cultural and historical ordering principle, structured and expressed institutionally and discursively, rather than as an objective reality. Much of late Tudor and early Stuart thought intellectually and psychologically signifies this dynamic process of redefinition. And while this increasingly more self-conscious process is not ultimately consummated in the seventeenth century, it does receive temporal expression in Hobbes' conception of man and his creation, the representative secular sovereign state. This serves as a foundation for a positive journey into a "modernity" structured and redeemed around a "historical" relationship between consciousness, order, and social reality.

Consciousness relates to social reality by limiting possibilities—this is order. It does this by mediating reality institutionally and discursively through social structures such as polity and language. Consciousness appears as the quality of perceiving order in things. And the history of consciousness is concomitantly a history of changes in modes of perception; it is thus a history of different "orders." Discourse tropologically reflects "the processes of consciousness [and] in fact underlay[s] and inform[s] all efforts of human beings to endow their world with meaning."[6] Consequently, consciousness orders discourse in the same way that discourse orders (structures) social reality. The epistemological and interpretive quality of ordering is one with the culturally and historically ordered nature of social reality. The history of consciousness is really consciousness in, or as, history. And order is the historically articulated relationship between consciousness and social reality. Indeed, the history of consciousness and of discourse reveals that consciousness reflects *history as order* and that the self-consciousness of consciousness defines *order as history*.[7]

A deeper understanding of the characteristics of the shift from the Tudor to the Hobbesian idea of order will help focus attention on the historical psychology of the origins of modern Western culture. Modernity may perhaps best be described as a process whereby man created the self as the historical and cultural medium for redemption. And while persuasive arguments have been made for medieval or eighteenth-century origins of modernity,[8] I believe that in the late sixteenth and early seventeenth centuries in England this process became objectified. The development of the self as a conscious gnostic[9] force for temporal secular salvation can be traced by an interpretive journey into Shakespeare's "vision," and by an analysis of English political and economic thought during this period.[10]

There are studies that identify the early seventeenth century as the formative period for modernity. In *The Revolution of the Saints*, Michael Walzer argues that radical Calvinistic Puritanism is related "not with modernity but with modernization, that is, with the process far more significantly than with the outcome."[11] Walzer sees the Puritans engaged in a battle against the traditional old order, alienated from the society that was representative of that order, and actively defining a new sense of order based upon the anxiety that necessarily accompanied such alienation.[12] Walzer does recognize that radical Puritanism can best be understood as a reaction formation, but he is wrong, I believe, to stress the conflict between traditional order and Puritanism as the dynamic conflict out of which modernity or modern political and personal conceptualizations arose. By posing such a conflict as basic to the process of modernization, Walzer devalues the secular aspects of modernization.[13]

Puritanism was only one of a number of alternative reactions that surfaced as the traditional institutional order waned. Theologians, philosophers, poets, and political theorists considered new alternative means to replace the traditional order.[14] And as was the case with Puritanism, often the creative and radical in these alternatives were characterized as much by a relation to emptiness and isolation as by an opposition to the traditional. Basic, then, to redefining meaning and order was the psychological problem caused by the absence of a well-defined source for identity. The common characteristic of each of the alternative means to order and identity was a dependence upon and a consciousness of self as the defining force in the new order. What was sought, as contemporary literature of all genres makes clear, was a means whereby the creative potential of the self could be refined and rationalized. Subsequently a new society-self relationship could be understood as the basis for personal and political order. Such a relationship was solidified by an increasingly active acceptance of secular attitudes in political and economic life. The concept and place of the secular sovereign state as the institutional foundation for social order and identity was facilitated by a more secular attitude toward life and by a general idea of social utility that acted as the necessary rationalizing mechanism for the creative self. In this way two anxieties were overcome: the anxiety caused by the dissolution of the old order which forced the conscious citizen to look within self for a way out, and the consequent anxiety felt by the alienated citizen as he tried to define security in self.

A study of radical Puritanism, even one that considers Puritan political theories concerning order, cannot identify the basic structures of the intellectual and psychological changes that dynamically altered the English concepts of society and order in the early seventeenth century. Radical Puritanism is one manifestation of the crisis in order that engulfed early modern England. But its response was, I believe, not the one that resolved this crisis.[15] It is rather to dramatic literature, as a way of interpreting the contemporary psychology of the meaning of the relationship between self and society, and to economic and political theory, as a way of analyzing this relationship in terms of institutionally objectified structures, that I will turn, in order to focus upon the kinetic process of English intellectual history in this period.[16]

The Tudor idea of order with its manifest concepts of correspondence is part of an eschatologically Christian view of the world.[17] Puritanism, while denying the traditional idea of order and correspondence, still defines a world view that is eschatologically oriented, that is, based upon a final transcendental truth. Hobbes' political order, however, posits a radical celebration of noneschatological existence.[18] There is no purpose, beyond natural man who exists as a mobile, motive creature, which predetermines social, personal, or political order.

The move from the Tudor idea of order to Hobbes' is an all-encompassing one. It is possible to identify the establishment of secular values in early modern England by analyzing the structural modifications that characterized the larger metamorphosis in cosmic vision as represented by this changed idea of order. Very simply, the basic difference in the Tudor and Hobbesian ideas of order concerns their relation to change. Tudor theorists maintained the late medieval view that order was natural and that change was a threat to natural order. As Pocock observed, "the character of European political thought towards the end of the medieval period . . . [indicates] a culture with a very strong bias towards believing that only the universal, the unchanging and consequently the timeless, was truly rational."[19] In this sense social meaning was a component of order, discursively structured and metaphorically reflected, through correspondence. Hobbes understood that a world in flux was natural and that order must be created to restrain what was natural. Now, order instituted socially constructed meaning metonymically through re-presentation.[20]

This transition in the idea of order marks a move from the idea of society to the idea of the state, or at least a metamorphosis in the concept of the state.[21] The locus of order shifts from the divine cosmos to the secular sovereign state. The idea of a state, in this context, suggests that social order is separate from divine, natural, or cosmic order. Social order, now, begins to incorporate rational utilitarian authority. And authority is no longer a matter of acquisition or definition; it is only a matter of exercise. Secular order then, redefines the social good. Society is no longer a transcendentally articulated reflection of something predefined, external, and beyond itself which orders existence hierarchically. It is now a nominal entity ordered by the sovereign state which is its own articulated representative.[22]

It is my contention that this transition in the idea of order signifies the emergence of the secular sovereign state as the dominant structure in society. The literatures of the age indicate sustained efforts to revitalize or to refashion familiar and rewarding social definitions of existence. Whether theorists openly advocated or intended change, which they rarely did, or whether conservative rhetoricians implored for a return to ancient custom and law, the thought and action of this period were inherently innovative and radical. The difficulty among even the most educated to accept change and the new value formations that informed the challenge to the old order insured that the intellectual history of the era would be wrought from conflicting, inimical sets of values, and that the history of change during these years would be a history of conservative revolutions.[23]

Contemporary social scientists of various persuasions have noted the applicability of T. S. Kuhn's theories about scientific revolutions for their own areas of study. The political world as well as the scientific world is paradigmatic, we are told.[24] In "normal politics," as in "normal science" there are basic "values and belief patterns," that sustain the paradigm and thus order the political world. Such commonplace values "go to the heart of the political order and give it identity and self definition."[25] Paradigms, then, can be considered "controlling concepts and theories [which] authoritatively indicate not merely the solutions to problems, but the kinds of problems which are to be conceptualized as requiring solution."[26] In this way revolutionary political change can best be understood as the "transformation of one conceptual scheme . . . for another."[27] Ideological revolutions occur, then, when "something has happened which necessitates a redefinition of the problems to be solved."[28]

In this same way the idea of order may be conceptualized paradigmatically—we can talk about "normal" or traditional values, culturally received ideas about self, society, and the cosmos which "meaningfully" reflect this paradigm, this ordering principle. And more important, we can investigate from this methodological perspective the cultural process of meaning redefinition. The transformation of the idea of order indicates a process of cultural production as well as cultural change. The social world is re-constructed, re-structured. In general terms, in the sixteenth-century order paradigmatically (discursively) structured a world of resemblance and similitude. In the following century, however, order structured the world by conceptualizing spatial and temporal relationships for different things.[29] An interpretation of this transformation should not presume that it necessarily reflects progress, evolution, or even continuity. Rather, if we investigate change from the perspective of paradigms and anomalies, we can focus attention on continuous as well as contiguous processes and thereby historically interpret the production of meaning.

It is necessary to guard against reductive interpretation when dealing with intellectual history. The vitality of the Elizabethan world is as much a historical commonplace as is the "Elizabethan World Picture." On one hand, the creative surging, exhilaration of the Sidneys, Drakes, Raleighs, and Shakespeares; and (on the same hand!) Elizabethan restraint, conservative politics, and reiterated harping on due degree, on correspondence, on immutability, and on order, the traditional order. Is there a paradox here? Is there a gap between social reality and conceptualization? Was the Elizabethan world absolutely traditional and conservative? Kuhn sharpens our vision:

> Often a new paradigm emerges, at least in embryo before a crisis has developed far or been explicitly recognized . . . in [some] cases . . . one can say only that a minor breakdown of the paradigm and the very first blurring of its rules . . . were sufficient to induce in someone a new way of looking . . . in other cases . . . considerable time elapses between the first consciousness of breakdown and the emergence of a new paradigm.[30]

Kuhn suggests that the usual response to an initial conceptual breakdown in the old paradigm is conservative. Concerted efforts are made to validate and to revalidate the traditional ordering of things. Problems are isolated and defined within the old system of identification. In this way the initial breakdown is overlooked, and the causes of disorder are disregarded. Only later will a new paradigm, a new ordering of things emerge, and only after sufficient battle.[31] "The decision to reject one paradigm is always simultaneously the decision to accept another, and the judgment leading to that decision involves the comparison of both paradigms with nature and with each other."[32]

It is not surprising that the best descriptions of the commonplaces of society and the traditional idea of order were written in the Elizabethan years when anomalous attitudes in science, religion, politics, and poetry began challenging the traditional commonwealth. Just as historical, secular, relativistic, and existential attitudes were most conspicuously elaborated when "modern" culture began to reflect ironically and self-consciously on its meaning, so too was the "Elizabethan World Picture" most clearly articulated when it was no longer taken for granted. These "modern" attitudes, I suggest, resulted from cultural meaning redefinition. They were first experienced as anomalies, then in some few cases articulated programmatically as alternatives. Though they did not become "paradigmatic" during the period I am investigating, I believe that in order to understand how meaning *and* social reality are cultural productions it is necessary to investigate—even to highlight—shared anomalous responses. Marlowe's canon is but the most conspicuous example of such seemingly radical expression. And, of course, there are those who seem aware of and ambivalent about both the new and the traditional conceptualizations of self and society. Shakespeare indicates the conflicts between alternative innovations as well as between the old and the new. Nothing is redefined without being refined.

The working out of these conflicts was an expression of social and historical psychology, of the cultural history of the self. Public, social values adjudicated against the self, or what seemed to be the initial expression of self-consciousness, self-interest. Literatures of all types as well as personal histories clearly point to the conflict between private desires and public interest as seminal to the cultural dynamism of the age. The question was a simple one. Were social values necessarily antagonistic to self-interest? Only in a society in which meaning was no longer taken for granted could such a question even be posed.[33]

Sociologists remind us that values are not concrete rules of conduct. They "may be simultaneously components of psychological processes, social interaction, and cultural patterning and storage."[34] While values motivate and shape goal-oriented behavior they also rationalize and justify past conduct. What are acceptable actions, what are socially remunerated aspirations at one time may be anachronistic posturings another time. Alternately, what is considered at one time a transgression in social behavior may develop into a prudent, practical commonplace.[35] Order coerces individual behavior so that meaning is both limiting and liberating. It is when order is most self-consciously articulated that it is perceived less as a cohesive and more as a coercive force. At such moments cultures suffer what sociologists refer to as anomie and individuals are often susceptible to alternative, "deviant" ways of knowing and doing. Difficulty arises when individuals' goals and/or their means to attain them are inimical to the accepted mores of society.

Arthur Lovejoy encourages "the historian of ideas to record the sequence of changes in what may be called the *approbata* and *admirata* in a given society."[36] These two are often contraventional. Remember Hamlet. Shakespeare dramatizes the confrontation between individual aesthetic admiration and customary, approved social mores. Think of Hamlet's considerable self-doubt and follow Lovejoy:

> There are . . . instances of admiration which are spontaneous essentially aesthetic responses in the individual to certain characteristics or modes of action in other men; and these responses may be out of accord with the ordinary moral criteria which the same individual accepts.[37]

How does social meaning change? How are the changes in the individual's consciousness of self and of "other" related to this redefinition of meaning? A study of the transition in the idea of order in Elizabethan and Jacobean England cannot be restricted to a history of ideas, no matter how perceptive such an analysis may be. Since order can be conceptualized as the perceived historical interpenetration of self-consciousness and social reality, a careful discussion of the relation between the idea of order and the redefinition of meaning requires, both as complement and as supplement, an analysis of the changing consciousness and conception of self during the period.[38]

We can glean profound insights into the struggle between the individual and society that intensifies during these years from personal histories, political debates, and especially from creative literatures. A brief introduction to an interpretation of Shakespeare, organized loosely with Kuhn's model in mind, can set a foundation here. The festive comedies such as *A Midsummer Night's Dream* and *As You Like It* and the *Richard II–Henry IV Parts I and II* sequence dramatize the dissolution of the traditional ceremonial bonds of society, and consequently, of the traditional self-society relationship from which the self gained identity from outer-defined social norms. Alternative value systems compete to define and justify behavior in *Troilus and Cressida, Julius Caesar*, and *Hamlet*. Slowly, but perceptibly, the self, the individual, begins to gain stature as a source of security and as a possible agent for responsibility. Yet while the traditional ordering of things is waning, the new alternatives are still undefined, unrefined. Aspirations associated with the self, with private demands, remained anathema to social, public goals in the eyes of most.

"Lacking a stable world of values, many individuals were inclined to take their own instincts and passions as the only guiding standards of behavior."[39] The specific historical accuracy of this general, vague assertion is less important than its suggestiveness about psychological confrontation. In *King Lear*, in *Macbeth*, and especially in *Coriolanus* security and identity are sought in the authentic isolated individual. The way out of the increasingly worsening conflict between private and public becomes the way in. The self becomes the final and total source of truth, of value, of life. Correct behavior is predicated upon the urgings of the self. Consequently, everything which is not self is untrue, valueless; and yet truth now becomes isolation. Cordelia and Coriolanus tragically turn to their selves as sources of strength against societies whose public truths no longer assure meaning, order, and redemption. For each the self is the only alternative, and the self equals truth. But truth is "nothing."

There was dramatized at this time a need to construct a new frame of reference for individual behavior that was based upon private, individual values.[40] This would necessitate a private as well as a social acceptance of the individual and of one's perception of oneself as an individual, as an actor in the world, responsible for it. Shakespeare's last plays confront this requirement directly. Prospero is the actor-creator who redefines meaning as he defines social harmony. Using powers that have been mellowed and refined in isolation (on the island), he responsibly reconciles all antagonisms and private-social oppositions. Although a pessimistic undertone pervades *The Tempest*, the encompassing feeling is one of regained order and security. This play may suggest a return to the less introspective, the less problematical mood of the earlier festive comedies. Indeed, here reconciliation derives from generational peace and marriage as it did in *As You Like It* and *A Midsummer Night's Dream*. Again at the end, a social order is celebrated from which the individual is not alienated; nor is he antagonistic toward it. The great difference, however, is that this is Prospero's order, self-consciously constructed and articulated. Prospero harnesses his own energies and creates. Now the individual, the self, defines a new self-society relationship. The social order neither subsumes the individual nor forces him into an isolation which is negation. Prospero's creativity negates negation. It focuses one's powers outward. It is a manifest expression of the individual's attempt to secure meaning, order, and finally redemption through the mediating self.

Prospero is an actor, but more than that he is a magus. Here, in the tradition of Cornelius Agrippa and of John Dee, Prospero the individual works in harmony with the forces of the cosmos. He directs what is; he works to actualize all potential sympathies.[41] The Renaissance magus operated in a cosmos that very closely resembled the medieval world picture. In this way, Elizabethans involved in intellectual Hermeticism found a new means of expression, a new force with which to seek order and harmony, which in nearly all manifestations appeared traditional and religious.[42] But the basic difference, as Prospero exemplifies, was undeniable. *Man* had changed. He was the operator, the actor, the manipulator now. And whether he chose to operate, as Prospero did, in a magical world, for religious and imperial ends, or as many others did in various secular activities, for vocational ends, he was responsible for his world and its order.[43]

A discussion of intellectual Hermeticism and Renaissance magic shows that some challenges to the old order appeared conspicuously traditional. Indeed, for many, some form of Renaissance magic remained a plausible, traditional, creative alternative both to the old order and to new radical alternatives, well into the third decade of the seventeenth century. Yet what makes even this religiously motivated alternative indicative of the intellectual and psychological changes that fashioned the challenge to the "Elizabethan World Picture" was the emphasis on action and operation that characterized it.

As Frances Yates explains, the desire to operate that characterized the magus led to the actualization and the sanctioning of the individual will. The magus "changed the will" and consequently dignified and justified individual actions.[44] The Hermetic tradition which underscored the religious-magic activities of the late sixteenth century could also be translated into political and intellectual action. The Hermetic emphasis on order, harmony, and eclecticism was a strong foundation on which to build an Eirenicist vision. The religious wars that plagued France, the Low Countries, and Germany in the last years of the sixteenth century were emblematic of the discord and disorder that religious dissension and an uncertainty about faith provoked. A new faith in religion and in magic would help secure a harmonious, orderly Christian Europe. Yates involves Sir Philip Sidney and Giordano Bruno in the late sixteenth century and Frederick V, Elector Palatine, king of Bohemia and Elizabeth Stuart in the seventeenth century in this religious mission. Whether or not there really was any specific "Hermetic" plan for a united Europe, the emphasis on a defined and created order that was associated with such ideas was crucial.[45]

The Hermetic experience also added new complexities to the commonplace harmonic microcosm-macrocosm sympathies that described the Renaissance English world. The fixed simplicity of the "Great Chain of Being" was naturally challenged by the energized world of the Hermeticist. Pico della Mirandola and the Florentine neoplatonists and the Venetian Franciscan, Francesco Giorgio, directed the intellectual world's attention to number as the key to all nature.[46] The new mathematical descriptions of the cosmos that gained currency in the seventeenth century were influenced by the intellectual excursions of Renaissance Hermeticism. Indeed, the emphasis on operating and on number that was clearly part of the Hermetic world facilitated the development of modern science. A "new thinking cap" as well as observation and experimentation produced modern science.[47] And again it was the willingness to involve the individual as doer in the larger world that led to a conception of a satisfactory and acceptable order for the universe.

Some Elizabethans and Jacobeans found a way out of the political, economic, and religious tensions of their times in the spiritual, intellectual, and magical world that was directly or indirectly influenced by the Hermetic tradition. This should not be difficult to understand. In one sense this was another atavistic, yet legitimate appeal to a traditional ordering of things. What made it a timely exercise, however, was its innovative qualities. Other less intellectual characteristics of this tradition are similar in many instances to the popular magic that was prevalent during these years.[48] In any case, the importance of this tradition cannot be overlooked. As an alternative means to an ordered and certain existence opted for by some of the most sensitive and intellectual people of the late sixteenth and early seventeenth centuries, it must be considered. How different was it from other alternative means to order that seemed plausible at the time, especially ones that shared common attributes with it as did the new science? Were its means directed, by men such as Marlowe's heroes, toward secular and temporal, rather than spiritual, goals?[49]

As private attitudes became public values in the process of providing the individual with psychological security, so too all other alternative means to order were concerned with social utility, the key to redefined social meaning. Utility secured self-in-society. The merchant taught how his profits were the commonweal's profits. The Parliament man, Puritan or not, who argued against the insecure succession or the royal prerogative, did so to strengthen the commonwealth. Francis Bacon advised James for England's good and catalogued knowledge for society's benefit. The Hermetic cause was no less socially conscious. Giordano Bruno understood that the ethic of social utility and public service served as foundation for a reformed society, the goal after which he strived. Elaborating traditional humanist educational values, he argued that all learning and invention were for the well-being of society.[50] In all of this can be seen evidence of the development that made magus-like individualism socially utilitarian-oriented.

In what follows I will illuminate what I have but briefly introduced here. This is intellectual history, "the attempt to write history of consciousness-in-general, rather than discrete histories of, say, politics, society, economic activity, philosophical thought or literary expression";[51] it is concerned primarily with meaning production and redefinition. Though much will be gleaned from political philosophy and popular pamphlets, it was the literature, of Shakespeare and Marlowe primarily, that motivated the questions that led to this book.

Rather than represent or negate a prescriptive ideology—an "Elizabethan World Picture"—their works accentuate the culturally contestatory process of meaning redefinition. And, it is conspicuously to indicate this process and experience that Chapter 2 focuses on Shakespeare's dramatic vision. Shakespeare dramatizes the problem of order, not the solution to this problem. "His historical and tragical worlds neither affirm nor deny Tudor visions of order and harmony; they concentrate, rather, on what is at stake in the mind's longing to behold, or to challenge, such visions."[52] They dramatize the continual cultural process that is social change, and allow us to appreciate that history is meaning and meaning is becoming. In so doing, they suggest the interpretive avenues down which we must pass if we are to offer a particularly historical answer to the generally articulated question posed nearly forty years ago by Francis Fergusson: "At what point in history, and by what process, was the clue to the vast system of Medieval analogies lost, the thread broken, and the way cleared for the centerless proliferations of modern culture?"[53]

Chapter 1 will establish the commonplace paradigm of order as it was ubiquitously described during the Elizabethan years. Then the idea of order as it was understood in the 1640s will be discussed. In both cases the relationship between the idea of order and conceptualizations of state, society, self, change, and immutability, and various other attributes of social meaning will be interpreted. Hence the old and the new paradigms will be put on stage. How and why did such a radical change come about?

Chapter 3 looks at anomalies that surfaced within Tudor order theorizing and considers some alternative postulates that began to develop in the later Elizabethan years. In Chapter 4 alternatives that surfaced in the first decades of the seventeenth century will be discussed. The conflict was no longer between the old paradigm and a substitute one, but between various possible new paradigms. It was during these years that the creative urges that conditioned the attack upon the old order were being refined. It was also during these years that secular and temporal attitudes in politics and economics began to challenge the taken-for-granted cultural mentality.

The last chapter listens to the voices that most articulately indicate a revolutionary idea of order. Filmer, Selden, and Parker will be interpreted from this perspective; so too will pamphlets on politics and society that glutted the market in the first years of the 1640s. But Thomas Hobbes will have the last word. It is he who most fully defines the secular sovereign state and the order inherent in it. Hobbes overcomes the conflicting alternatives, whether they represent transcendental or existential truths. The conflicts that characterized this historical age were so creative because basically they were internecine. For this reason, each alternative, as different as it seems, may be understood to express similar historical psychological experience, and each then suggests an absolute resolution to the cultural search for meaning redefinition.[54] It is such a vision of complementarity that Shakespeare offers. And it is basically a tragic vision because what is really being sought is not absolute truth, as being, but order and meaning as becoming.

Hobbes overrides the alternatives and denies complementarity as well as any singular truth. He makes natural variety the sufficient reason for order and he projects this order outside of himself, onto society. Hobbes' order retreats from the Shakespearean vision of multiple possibilities without positing instead, as Puritanism did, one possibility, one truth, only. Only an artificially contrived order, a force of voluntary repression, could stop the internecine struggles that were so manifestly expressed in England from 1640 to 1660.

CHAPTER 1

From Divine Cosmos to Sovereign State

> The heavens themselves the planets and this centre,
> Observe degree, priority and place,
> Insisture, course, proportion, season, form,
> Office, and custom, in all line of order.
> Shakespeare, *Troilus and Cressida*
> (I. iii. 87–90)

How many historians, political scientists, and literary critics have cited Ulysses' famous "degree" speech from *Troilus and Cressida* as a perfect example of the Tudor idea of order and of the taken-for-granted Elizabethan meaning of the world?

> Take but degree away, untune that string,
> And hark what discord follows! each thing meets
> In mere oppugnancy: . . .
> (I. iii. 109–111)

Scholars have traditionally cited Ulysses' words to illustrate the staunchly conservative nature of the late sixteenth century.[1] It should be clear, however, that the ordered universe Ulysses describes and the world he knows and willingly and gleefully participates in are vastly dissimilar. Ulysses is aware of this. In this speech he seeks alternative ways to bring a new order and consequently a fresh success to the Greeks. He and Nestor concoct a pragmatic brew simmered by the rhetoric of order and degree.

I am not suggesting that there really was not an idea of order that informed Tudor commonplaces and behavior. Nor am I suggesting that Ulysses was merely a deceiver, that he had no more regard for order than did Tamburlaine, and that his recourse to the rhetoric of order, harmony, and degree was but political strategy. The concepts of order, degree, and harmony were meaningful for Ulysses and his Greeks as they were for many Tudors. But they were meaningful because they helped explain and define social meaning as well as individual actions and relationships. While there was a clear Tudor idea of order which can be described (and often is) and which can be captured in a key poetic line or two, the Tudor idea of society and of the individual in society which derived from this idea of order were more complicated. Until the end of the sixteenth century the idea of order remained a dynamic concept capable of explaining various changing social attitudes without necessitating a self-conscious value reassessment.[2] This traditional idea of order which described the "Elizabethan World Picture," rather than restricting the English in the fixed, static, unchanging world that it metaphorically presented, allowed them the security to roam somewhat frenetically across the horizons of human possibilities.

Troilus and Cressida is about values and valuation; or more precisely, it is about the lack of certain, secure criteria for evaluation. Are things and acts valuable in themselves and for themselves or does everything derive its purpose and meaning from an external source? Is Helen worthy and thus Greeks and Trojans fight for her, or is Helen worthy because Greeks and Trojans fight for her? Ulysses and the rest have to explain and to justify their world and their actions in it. And although the gap between social reality and social meaning continues to widen, until meaning is redefined and represents a new perception of social reality, the old concepts of order and degree continue to exist and function as explanations.

In the late sixteenth and the early seventeenth centuries the traditional Tudor idea of order continued to explain social reality, albeit less convincingly as the years passed. But it no longer limited this world so definitively. Its capacity to serve as a foundation for contrary attitudes about society during the Tudor years contributed profoundly to its political value in these economically and socially unstable years.

Social theory during the reign of the eighth Henry still owed much to medieval ideas. The idea of a mixed government held sway. Indeed, some Henrician pamphleteers even hinted that the people could resist a tyrannical prince. The Rolls of Parliament at the deposition of Richard II epitomized the view that the people are absolved of their oath to the king if he violates the coronation oath. Thomas Starkey offers a good starting point. He argued in the *Dialogue* that no country can be well governed unless the prince is elected because hereditary kings are rarely worthy.[3] In the *Exhortation to Unity and Obedience* his order theorizing was more characteristic of medieval and Renaissance thought than of later Tudor thought. In all nations' policies, said Starkey, "the worde of God must be of chiefe authoritte. . . . Wherefore if anythynge be decreed contrary to that, it must be bitterly abrogate and boldly disobeyed with all constancy."[4] Even in 1559 when Elizabeth took the throne amidst concern about the danger of a woman monarch, John Alymer used medieval arguments to defend his queen. It is less dangerous to have female rule in England than anywhere else, he argued, because "the regiment of Englande is not a mere Monarchie as some for lack of consideracion thinke, nore a meere Oligarchie, nor Democrate, but a rule mixte of all these wherein ech one of these have or should have like authorite."[5]

The idea of order, however, could as easily justify a more direct Tudor "Cult of Authority."[6] The necessity of internal peace, sharpened by memories of recent civil war and by an unstable and inflationary economy, the defense of royal supremacy, and the defense of England against Rome and Spain, were three pragmatic reasons for the Tudor concept of authority. In the *Homilies*, the officially authorized Tudor word spread from the pulpit and shaped the popular consciousness. It is in the *Homilies* that the precise language of authority and order, a language repeated by theorists and politicians, is most evident. It is unlawful for subjects or inferiors to stand against superior powers. Man cannot in any case rebel against rulers. "Rebellion . . . [is] . . . the first and the greatest, and the very root of all other sins." To resist authority, even tyrannical authority, is to resist God.[7] By 1587 the threat from Spain and Rome was at its height, and anxiety about the succession rampant. Richard Crompton reiterated the by then commonplace exhortation to order and obedience and maintained also that it is not lawful "for the Subject, to enter into the examinations of causes or matters appertayning to ye Prince and soveraigne governor."[8] Questions of prerogative and sovereignty grew alongside the developing "Cult of Authority."[9]

It is evident, then, that the Tudor idea of order, while defining a static, immutable universe, was a dynamic concept capable of informing various social and political conceptions. In general, the Tudor idea of order described a divine cosmos very similar to the one fashioned by the traditional "Great Chain of Being."[10] All creation was by degree situated in a divinely ordained chain of existence. "Almighty God hath created and appointed all things in heaven, earth and waters in a most excellent and perfect order."[11] In this chain, man's place, as Hamlet reminds us, lay between the angels and the beasts. There was order in the heavens, in nature, in the seasons, in society and in man where could be found "soul, heart, mind, memory, understanding, reason, speech, withall and singular corporal members of his body in a profitable, necessary and pleasant order."[12] Order provided perfect correspondence between all parts of the chain and perfect harmony between each level and between all the degrees within each level. So the hierarchical structure of the cosmos, permeating the individual levels of the chain, too, meant that some men, princes for example, were higher than others—and all was but a matter of degree.

Order also meant fixity, constancy, immutability. Change was the greatest of all enemies. Tudor thinkers viewed time and history cyclically. Flux meant chaos and chaos meant a return to pre-Creation disorder. Early in the century, Sir Thomas Elyot, a Henrician humanist, set the model for statements about disorder and chaos. Much as Ulysses would ask seventy years later, Elyot questioned: "Moreover take away order from all things, what should then remain?" His answer: "chaos . . . where there is any lack of order needs must be perpetual conflict."[13] His reasoning was clear. God has made everything—all his works—full of degrees and estates, Elyot informed his readers. The Bible proves that there is order in the heavens. There are the four natural elements, and of course, there are all of God's creatures who exist within order and degree; "so that in everything is order, and without order may be nothing stable or permanent."[14]

Tudor theorists maintained that order was natural, indeed divine. Any social or political order that existed reflected natural, godly defined order, and this reflection corresponded to natural order. It did not represent it. "Nature is nothynge else but God him selfe, or a divine order spred throughout the whole world and ingrafte in every part of it," John Aylmer affirmed.[15] Consequently, security and immutability were maintained through this natural order. "For God so careth for the preservacion of this godly and comely frame of his, the world:

that he will not leave it without means of order, whereby it may contynew."[16] Otherwise what chance would there be to preserve man from chaos? Man alone could not secure order although the magistrate had the distinction of maintaining order in the civil state. But still, in dealing with kings, "we should rather fixe our eyes upon their office, which is gods: then upon their person whiche is mans."[17] Consequently, even the order which secured the temporal, civil world was fashioned outside of man and thus "if this order were not in nature al thynges woulde growe to confusion."[18]

Everything and everybody had one purpose, and that was God's purpose. Order prevented mankind from slipping into chaos; but it was also God's desire that all should be ordered. Starkey described the earthly world as "so knytte by dewe proportion in a certain equalitie."[19] An ordered civil life most perfectly reflected the divine telos and such correspondent order must be assured in "mans actes and fashion of lyvinge here in policie."[20] Man need not devise his own plan for wise policy or orderly living. God had ordained all things, including the hierarchy of "one man over another: that some ruling and some obeing, concorde and tranquilite might contynew."[21] A contemporary of Elyot's and Starkey's, Richard Morison, reminded the readers of *A Remedy For Sedition* that hierarchy was not "arm in arm, but the one before, the other behynde." The wise should rule and the strong obey because "ahandful of witte, to be moch more worth than a horslode of strengthe."[22]

Most agreed that civil order and politic rule were necessary, and that by following one's duty, whatever one's degree, such order could be maintained. Throughout the century this theme was commonly sounded. John Ponet stressed proportion and obedience. As the body is kept in proportion by sinews, he argued using typical Tudor tropology, "so is every comū wealthe kept ād maītaned in good order by Obedience."[23] In contrast to later political theorists who defined extremes against the middle way, Ponet noted that balance and proportion were the keys to true order. Drawing from contemporary examples, Ponet singled out the Anabaptists for their lack of obedience and order and the Catholics for their overorderliness. George Whetstone harped on this theme too, while writing in defense of orderly cities. Plowmen and inferior people were "the feete, which must run at the comaundement of every other member . . . where this Concorde is, peace and prosperytie floorysheth in their Citties."[24] An ordered civil society was, then, a due correspondent reflection of God's divine order and security against discord and chaos. Order was assured by each person following proportion and fulfilling the duty inherent in his degree. Or as Aylmer put it in 1559, "Where good concorde, and brotherlye unitie, where loyaltie and obedience is; there must needs bee a sure state."[25]

In attempting to affirm and to reaffirm the ordered society, Tudor thinkers expressed the commonplaces of their age which were informed by the idea of divine order. The king's behavior was a source of concern.[26] Without exception, the belief in the necessity of good council was accepted. John Aylmer argued that Queen Mary had been deceived by her churchmen, evil counselors. "Spirituall men shuld not medle with policies," he maintained.[27] His advice to his monarch was to "look not to the person but to the reason," in hearing counsel.[28] All this "advice" Aylmer documented with biblical quotations and with illustrations from nature.[29] *Gorboduc* came right to the point. The first cause of chaos is "when kings will not consent to grave advice."[30] The effects of bad counsel were dramatized in the persons of Hermon and Tyndor, the respective "parasites" advising the young princes, Ferrex and Porrex. "As befel in the two brethren, Ferrex and Porrex, who refusing the wholesome advice of grave counsellors, credited these young parasites, and brought to themselves death and destruction thereby."[31]

Other commonplaces explained the nature of the king, the requirements for kingship, the different types of rule and the different kinds of law.[32] Social structure as expressed publicly and privately made manifest the divine order. Proper degree prevailed in the body, in the family, and correspondingly, in the "body politic," too. For Starkey, "the thyng whych ys resemblyd to the soule ys cyvyle order and polytyke law."[33] Just as a man's head and body exist together, orderly, so too does the head and body of the realm, the Parliament. Sir Thomas Smith understood the concept of king in Parliament when he explained that the prince was "the head, life and governor of this common wealth," while the Parliament was "the whole head and body of the realm of England."[34] Everyone was represented in Parliament, Smith commented, "from the Prince to the lowest person of England."[35] It should be noted again that this representation was not understood as substitution. Men did not negate their own wills in the hope of having them represented legislatively. Parliament did not yet "act for" the country by willing law, rather it revealed law; it was a court. Representation at this time meant that the Parliament corresponded to the individual; it was the "body politic."[36]

Order and degree were at the heart of Tudor concepts of society and the commonwealth. The sixteenth-century Englishman recognized theories of society, not of state.[37] The state was subservient to a society which revealed God's laws and order. In the years following the Henrician reforms when the Church lost its monopoly on morality and came increasingly under the power of the state, the Tudor mind still felt that "Church and State were in general simply different aspects of a unitary society."[38] Sir Thomas Elyot began his famous *The Book named The Governor* with a humanist description of society. "A public weal is a body living, . . . made of sundry estates and degrees of men, which is disposed by the order of equity and governed by the rule and moderation of reason."[39] The commonwealth was ordered as was divine nature and it was maintained by duty and degree. John Aylmer stretched the analogy between divine nature and the commonwealth until it tautologically assured the reasonableness of accepting Elizabeth's rule. While agreeing that "whatever is agaynste nature the same in a common wealth is not tollerable," he argued that, likewise, whatever preserves a commonwealth has to be natural.[40]

If a commonwealth was truly ordered by degree, each member in it, from prince to pauper, had his duty to perform. The famous Elizabethan "Homily against Disobedience and Wilful Rebellion," which J. W. Allen suggests is "the completest expression of what might be called the Tudor theory of the duty of subjects in a commonwealth," addressed those who contemplate stepping beyond their given estate and thus rebelling against the society.[41] It was vanity to claim that rebellion may redress grievances and improve the commonwealth, "rebellion being . . . the greatest ruin and destruction of all commonwealths that may be possible." Rebellion is caused simply by ambition and ignorance. Ignorance is the lack of knowledge of God's revealed will which teaches obedience and the abhorrence of rebellion; and ambition is "the unlawful and restless desire in men to be of higher estate than God hath given or appointed unto them."[42]

In the introduction to the translation of a French political tract, dedicated to Sir Francis Walsingham, Ægremont Ratcliffe immediately assured his reader that "there is nothing more decent, commendable, or yet more beneficial to man, then to be contented, and constantly stande to his calling: without coveting, as often as his fond affection shall egg him, to be other than he is."[43] Elizabeth's Puritan secretary probably knew his calling well enough. Whether he was ever content to stay within the limits defined by his position is questionable. If he read the pamphlet dedicated to him, Walsingham received a succinct education in Elizabethan political and social commonplaces. Ratcliffe eloquently conveyed the same message as did the *Homilies* as he focused his attention upon corruption, change, and social disorder. There was one primary cause for these evils, he taught: the "ignoring of vocation: I meane, that men doe not know or consider themselves, to be but particular members of an universall bodie: and that they . . . be called & appointed eche one in his degree . . . alloted unto him, to keep himselfe sufficiently occupied . . . in all vertue, to the common release . . . of the universal Politique bodie and societie of all men in generall."[44]

There was a positive side to duty and degree, too, for men of all ranks. Tudor theorists knew that the king served God's laws just as society did. The *Homilies* assured all that "the high power and authority of kings, with their making of laws, judgements and offices, are the ordinances, not of men, but of God."[45] The king had proportionately a greater duty to order and society than did anyone else because of his higher degree. If the king acted unnaturally, without regard to his position as magistrate and father of the commonwealth, he threatened the entire cosmic order as well as the security of his society. No one witnessing *King Lear* could have been surprised to see Lear's violent behavior accompanied by corresponding climatic violence. The king had to rule wisely and justly, then, and had to listen to "grave advice." But he was not singled out. Other wise and highborn men were expected to serve the commonwealth, too. Even such a didactic piece as *Gorboduc* was too frivolous, too personal a work for such a man as Thomas Sackville. Competent young poet that he was, he still gave up his private literary world to become a high-serving minister to his cousin, Elizabeth.[46] Well-born and intelligent men shouldered a heavy burden. They were expected both to serve, and to serve well. For as Thomas Starkey warned, disorder was caused by "wyse and polytyke men, whych flye from offyce and authoryte."[47]

So, of course, everyone had a determined part to play in maintaining the social order. Richard Morison explained that "a common welthe is then welthy and worthy his name, when everyone is content with his degree; gladde to do that, that he may lawfully doo, gladder to do that, which he seeth shall be for the quietness of the realme, all be it his private profit biddeth hym doo the contrary."[48] There was no place for private desires or occupations. Fulfillment came from maximizing the potential in one's public position for the common good. Although order was natural, and man, naturally, was thus inclined to order, he had to help order reveal itself. For this reason civil law, customs, and various ceremonies resulted from the "ayd and dylygence of man." Without such aid, nature "wyl soone be oppressyd and corrupt."[49]

Tudor theorists and politicians believed that there was a fundamental law of nature through which reason could be interpreted or declared. All human law, including the king's law, was subject to this law of nature.[50] As Richard Hooker said late in the sixteenth century, "where the king doth guide the state, and the law the king, that commonwealth is like an harp or melodious instrument, the strings whereof are tuned and handled all by one."[51] Even Henry VIII's and Cromwell's use of statute law to reform the English church and English church-state relations did not alter Tudor conceptions about laws. Parliament was still considered the high court of the realm which revealed and interpreted law, but did not make it. Baumer concludes that "it was still the universal conviction in the sixteenth century that statute law was derivative of, and subordinate to, that higher law which God had made manifest to man through his reason."[52]

Because civil law was not "made" by man, Tudor theorists argued that statute law could not authorize anything contravening natural law. The idea of order and correspondence substantiated the belief that the laws of the realm merely promulgated divine law. St. German averred that "it cannot be thought that a statute that is made by authority of the whole realm, as well of the king and the lords spiritual and temporal, as of all the commons, will recite a thing against the truth."[53]

Being such forces of revelation, laws were valuable reminders of order, and powerful influences to order. John Ponet asserted that laws were made to maintain policy that reflected God's purpose to man.[54] In 1587 Richard Crompton, writing to dissuade conspiracy and treason and to remind subjects of their duties, declared that "the Lawe is the perfect reason, fearing men from evill and wicked actes, [and] forcing them to the studdie of vertue and good things."[55]

Similarly, ceremonies continued to involve the subject in the natural and traditional concept of an ordered society, too. The Tudor, and especially the Elizabethan, world was very fond of ceremony and pageantry. The ceremonial emphasis of the Tudor monarchy has been well discussed.[56] Shakespeare's comedies and histories make full value of the emotional and psychological security afforded by ceremony. Most non-Puritan theorists defended ceremony as a positive means of emphasizing tradition. While ceremony was traditionally religious in nature, after the Reformation it "became above all an inducement to political order."[57] Thomas Cranmer openly advocated this more social use for ceremony, which he saw "as a commodity . . . for a good order and quietness."[58] Tudor writers, theorists, monarchs, social leaders, and churchmen all understood the value of ceremony for social order. Law induced order by revealing the law of nature; ceremony provided stability and security by celebrating the structure and relationships of the social world.

Within the fixed, immutable cosmos that divine order described, man's place was secure. "The nature and condition of man, wherein he is less than God Almighty, and excelling notwithstanding all other creatures in earth, is called humanity."[59] There were three qualities that together made humanity: benevolence, beneficence, liberality.[60] Not merely was man's place in the divine chain secure, but his nature was certain, too. Man was formed, as was society, or the stars, from certain qualities that the divine order described. The sixteenth-century Englishman had no psychological conception of the self as an individual defined by and within itself. Individuality existed, of course, but it was marked by one's specifically arranged place in the degree-oriented society. "There is not one, who . . . is not secretly by the unspeakable providence of God, called to some vocation: that is to say, to one manner of living, or other."[61]

Man's actions and his purpose for acting were totally "other"-oriented. It was not that man represented society or the larger order of earth. His existence was totally subsumed within it. Man was a correspondent agent in the divine cosmic order with no accepted personal motivations. Indeed, it was foolish to act upon any private motivation in hope of securing certain ends. "We be not our own but his [God's]: nor have our faites in our hands to ordein what we liste, but must as waxe yelde to his wurkinge."[62] "Burden of responsibility" was rarely articulated in Tudor England. An individual could affect another part of the divine order, of course, but it was impossible to change the arrangement, the relations, or the general structure of the divine order or any particular part of it. Man had his place and his duty in this order and his sense of himself was therein defined. Such an understanding of self demanded a private-public relationship that was securely public-oriented.

Nearly as often as Tudor citizens heard exhortations for order and against rebellion they heard inducements to forget private desires and act only for the public good. Nowhere in Tudor theorizing was there the suggestion that what was done for private gain could also be profitable to the public realm. The individual experienced neither isolation nor alienation. One's place in the divine order and one's duty to that order were incorporated socially and created a strong sense of belonging and security. This commonplace relationship between private and public acted as the cement for the Tudor social order until the last years of the century.

Elyot and Starkey constantly reiterated this theme. Elyot reminded magistrates to be "openly published," to refrain from private life, and to be always public men.[63] Starkey argued the traditional commonplace. Each man had his place and duty by nature. Man's actions have but one end here in life: the civil order of the commonwealth without regard of singular pleasures and profit.[64] The public good should be the end of all "cogytatyonys, and carys."[65]

The private-public relationship was so defined that little sense of accomplishment could be achieved outside of it. Attempts to neglect the public good could lead to frustration and even to social ostracism. Anyone who was drawn to his own pleasures and quietness and who did not use his abilities for the public good, Starkey said, "dowth manyfest wrong to hys cuntrey *and* frendys, *and* ys playn vniust *and* ful of iniquyte."[66] Starkey even questioned the Renaissance-Platonist ideal of contemplation. Was it socially responsible? Could it be personally rewarding in a society that defined success and fulfillment so publicly? One answer Starkey offered was that "lytyle avayleth vertue that is not publyschyd abrode."[67] In this way Tudor theory protected the individual against the claims of its own authenticity and self-sufficiency and simultaneously protected social order and meaning. "'Self', one's inner identity . . . was seen more in terms of the honor and reputation of one's role in life than as something generated out of ego."[68]

Shakespeare's Coriolanus and Achilles dramatize the psychological truth of Starkey's statement.[69]

Although during the second half of the sixteenth century many theorists noticed that, indeed, their contemporaries were becoming less publicly motivated, these same theorists echoed the traditional commonplaces regarding public-private relations. It should not surprise us that in the Elizabethan years when so much that was fresh and creative surfaced and made evaluation somewhat tenuous, the established commonplaces were resoundingly articulated. John Ponet asserted that "next unto God men ought to love their countrey, and the hole common wealthe before any membre of it."[70] Man's duty was still understood as a social, godly, one. Ambition was vilified. God had appointed man a "distinct charge, not to be exercised to his own particular, but to the reliefe & common maintenance of the universal bodie."[71]

The preface to the 1559 edition of the *Mirror for Magistrates* poetically expressed the close relationship between private and public. Ambitious men ("prollers for power and gayne") "seeke not for offices to help other, for which cause offices are ordayned, but with the undoing of other, to pranke up themselves . . . they seke only their commodity and ease."[72] The author drew the strictest relationship. The people of the commonwealth are good, he argued, when there are no bad leaders and the government is duly ministered.[73] The private is totally an extension of the public. Individual action is judged only in relationship to social order. Private behavior is understood only as a proper or an important manifestation of a publicly defined ethic.

George Whetstone's discussion of love, in *Aurelia*, exemplified this. Not content to accept the Platonic references to sacred and profane love, Whetstone acknowledged that love is separate from all desires of the flesh. "Love is a simple devine vertue, and hath his being in the soule, whose motions are heavenly."[74] Love can be manifested in zeal for God, in duty to country, in obedience for parents, and in affection for friends.[75] It cannot be found in any private desires. Indeed, there were no truly private manifestations of love. Whetstone accepted marriage only because it ceremonially symbolized nature's order.[76]

The private-public relationship that derived from the Tudor idea of order necessitated a severe repression of self-expression and potential. Sackville behaved predictably when he quit his poetry in favor of public service. Elyot suggested that "gents" should not repress the inclination to "paint with a pen" but should practice this "in vacant times from other more serious learning."[77] There was little consideration of the self-fulfillment that the poet or the painter achieved.[78] Poetry or painting or music did not primarily serve the artist or the world of art; these activities had purpose only within the context of the public good. Art's final cause, according to Elyot, was service to the realm. A captain who could paint could also draw good maps of enemy camps.[79] A nobleman may have knowledge of music, but he may not play often. He may use his knowledge "secretly, for the refreshing of his wit," but finally, of course, the study of music increases his understanding of harmony; in this way, music, too, served the social order by expressing the divine concord in the world.

Tudor theorists often evaluated actions, desires, vocations, and pastimes in this manner. The idea of order insured that all proper activities revealed the intricate harmonies of the cosmos, and thus aided the social order. Consequently, they existed for just that reason. By using the tautological arguments available in such a fixed, immutable order it was easy to prove that if an activity was not purposive in this sense, it was not a proper activity. Nowhere was any activity or benefit evaluated in and of itself.[80] Elyot's evaluations suggest the late medieval practice described by Huizinga of attributing allegorical meaning to all particular entities and thereby crystallizing all things into an ordered existence identifiable as divine. Dancing together, though objectionable to certain divines, Elyot suggested, is a "concord" of the special different qualities of man and woman.[81] It is reconciliation, as is marriage, and thus has ceremonial purpose. For Elyot, dancing appeared as a form of social comedy in the best Shakespearean sense of that term. Dancing manifests all the best qualities of orderly and noble life. It is an introduction to prudence, "the first moral virtue."[82] It teaches maturity and balance, and it is an exercise that proves the value of living without superfluity or indigence. Indeed, each motion (and Elyot numbers many particular dance motions) has a particular purposive quality; the third motion, for example, reflects providence and industry.[83]

Starkey's *Dialogue* corroborated Elyot's book on these questions of evaluation. In searching for "that thyng whych ys the welth of every particular man,"[84] Starkey arrived at this conclusion: all attributes—health, riches, virtue—are to be used "to the end by nature *and* reason appoynted."[85] No particular asset has any inherent value, and all value is derived from divine nature. Consequently, that which facilitates man's divine purpose is beneficial to man.

But what is man's purpose, and how is it known? Tudor theorists stated that man performed a duty to the commonwealth and that this duty was derived from, and helped maintain, order. But there were larger, more religious ends for which his social duty was only a significant part. Such an end was the knowledge "of God, of nature, and of al the workys therof."[86] Certainly a knowledge of God's works was facilitated by the order in the immutable cosmos, and man understood order through law and ceremony. Elyot defended ceremony because "we be man and not angels, wherefore we know nothing but by outward significations."[87]

They knew the world as full and immutable and, of course, orderly. They were sure, too, that good things and virtues were defined by God in natural law and not by human opinion.[88] "Suche thynges are good, not as appeareth to mannes corrupt reason, but such as be by goddes owne worde defyned, by the whiche rule only we muste examyne what thing is good with ryghte judgemente."[89] Of course there were customs, practices, even laws that were not defined by God's word and were consequently unnecessary for salvation. It was useful, however, to evaluate them. Tudor commonplace regarded such practices as "indifferent." Such indifferent customs or laws had no intrinsic value. Only the fact that they were established by authority made them good. But that was enough because authority bound the people to them "ye by the vertue of goddes owne worde."[90] Indeed, any custom or law duly authorized was good and had to be obeyed.

Because the world was naturally ordered, the individual soul was correspondingly ordered. A virtuous soul was an ordered soul; a corrupt soul was disordered or diseased.[91] In this way Tudor psychology understood that disorderly behavior in an individual was a perversion of what was natural and good just as a disordered society was a perversion of an ordered society. Disorder was unnatural. It was merely a negation of what was good and natural and had no definable existence of its own. Intemperance and confusion, Robert Mason explained, was reason engraved with lust and concupiscence.[92]

Virtue was an ordering of the individual that was in no way dependent upon the individual's behavior. On the contrary, virtue directed correct behavior. Action and behavior simply revealed virtue or corruption, order or disease. Consequently, an individual was not defined by his actions but by the full, fixed order of which he was a correspondent part and which his actions revealed. It is understandable, then, that a usurer would be held in the lowest esteem by Tudor society. His actions showed a discontent with "the things of Gods creation according to their own values." Usury attempted to "adde further provisions and means than ever was ordained of god . . . it may be concluded that the raysing of increase upon bare money, is unnaturall: . . . because God hath left nothing unprovided that is requisite for the use of man."[93]

As Elyot said, "we know nothing but by outward signification." Knowledge was the knowledge of what existed,[94] and it was "the maidservant of decorum."[95] Understanding for sixteenth-century Englishmen, then, required a fully determinate and unchanging cosmos that was constantly revealing itself and was consistently knowable, and "decorum assumed that a suitable response existed for every conceivable situation and that it was possible . . . to catalogue, learn and apply the formula."[96] Tudor consciousness related to social reality by limiting possibilities. The idea of order that resulted from this relationship epistemologically and tropologically secured the social and psychological foundations of Tudor life and defined social meaning. Meaning was fully defined outside of the self, and the individual was correspondingly incorporated into this other. The sixteenth-century rhetorical handbooks and manuals of decorum supported this. They were models for "the formation of an artificial identity,"[97] an identity, by the way, that was experiencing the mounting pressure of institutionalization.

"Logique is nothing els but judgement, and finding out of thynges, so . . . judgement and finding out of thynges, is nothing els, but logique it self," said Sir Thomas Wilson.[98] Such "logic" was certain and comforting. It directed thought outward, not inward. It located meaning and order outside of man and claimed they could be ascertained. And it focused attention on the importance of understanding and reason, not on man's introspective designs. Represented as decorum, such logic limited the self's possibilities, and self-consciously reaffirmed the taken for granted, while it minimized (ordered) emotional liberty.[99]

For most Tudor theorists "right reason" was the "monarch" of knowledge. It taught men how to govern, and how to obey. It revealed order and government rather than horror and confusion.[100]

"Right and true reason . . . hath a place above all earthly and corruptable things."[101] Reason did not result from man's thought, but rather, was an innate virtue unto itself, a divine quality which directed behavior. While many order theorists lamented how "corrupted" reason had become in men—"none can doubt if they behold themselves truly what they are, in respect of what they should be"—they constantly sounded the close relationship between reason and order.[102]

Thomas Starkey gave an account of the genesis of society, a rarity in sixteenth-century English theorizing. Originally men lived in forests without reason or rule. They were governed by fantasy. Then by the gift of reason, men of great wit and policy perceived "the excellent nature and dygnyte of man." Following reason, they persuaded the others to follow nature, order, and civility. They devised laws that helped reflect their excellent nature.[103] Reason was the source of all civil law. God created man alone, said Starkey, with "a sparkyl of his owne dyvynyte,—that ys to say, right reason,—whereby he schold governe hymself in cyvyle lyfe *and* gud pollycy."[104] And Elyot agreed: "It may not be denied that all laws be founded on the deepest part of reason."[105]

Tudor, and especially Elizabethan, commonplaces and ethics were based upon the conciliation of the two highest human faculties, understanding and will.[106] Contemporary psychology divided the body into three areas, the highest being the seat of reason which directed human action.[107] Undirected will led to discord and chaos. Order and degree, the prerequisites for a public weal, revealed God's disposal of the "influence of understanding."[108] Good counsel, that ubiquitous Tudor commonplace, was right, good, and honest, said Elyot, because "that which is right wise is brought in by reason . . . [and] nothing is right that is not ordered by reason."[109] Reason was right reason when it motivated rational behavior and all rational behavior was naturally motivated by reason, not self-consciously by individual passions or interests. In modern terminology, Tudor social psychology was superego-oriented. The individual ego and id were restricted. The more the ego restrained the id, the more it resembled the superego and the more it appropriated "right reason" as its own venue.

Continence and temperance were much applauded virtues. Continent and temperate man "delighteth in nothing contrary to reason."[110] The continent man resisted unreasoned desires because "continence is a virtue which keepeth the pleasant appetite of man under the yoke of reason,"[111] and the temperate man followed reason because he "desireth the thing which he ought to desire."[112] It is clear that Tudor ethics depended upon repression, and that reason was the repressing agent. As long as reason was understood to be a divine virtue inspiring men's behavior, Tudor ethics strengthened the idea of order. When, however, reason became more closely identified with a human faculty, as it did as the century progressed, internal conflict ensued. Tillyard maintained that "the battle between Reason and Passion . . . was peculiarly vehement in the age of Elizabeth."[113]

Because of the premium placed upon such "decorous" behavior, individual will and passions were constantly denigrated by Tudor writers. Self-conscious decision making which abided by personal desires was considered unnatural. Tudor theorists did not recognize the value of introspection as a means to truth and good behavior as Hobbes would do a few years later. Elyot claimed that a man was "intemperate which by his own election is led."[114] In this sense intemperate meant unnatural and diseased. If reason is "overrunne," said Starkey, man slips from felicity into "infinite misery and wretchedness."[115] All the change and disorder that plagued the commonwealth, "all this fault cometh nowhere els, but of the corruption, and deprevation of man's wil and judgement from the which there proceedeth a blind confidence and presumption of himself, which is the spring of al errour."[116]

No matter what else Tudor men and women feared, "mutability" was public enemy number one. The idea of order assured changelessness, and for a society that believed in inevitable historical cycles such an overwhelming cosmological structure was a psychological imperative. Anything that was new was dangerous. John Aylmer argued that the controversy surrounding Elizabeth's ability to rule introduced new questions. All new things challenged the domestic order, he warned.[117] The *Homilies*, political theory, legal theory, indeed all Tudor thought preached against change and mutability. "Nothing is more proper" than stability and constancy; mutability is a great detriment, taught Sir Thomas Elyot.[118] "Constancy is as proper unto a man as is reason . . . it were better to have a constant enemy than an inconstant friend."[119]

Though the heavens were perfect and unchanging, no one could deny that the earth, man's home, was plagued by changes of all sorts. While sixteenth-century Englishmen could take care not to provoke rebellion or any other manifest social change, they could not escape the simple workings of time. Man's world was an imperfect reflection of the divine order. The poets, especially, confronted the reality of time and decay. George Cavendish wrote his *Metrical Visions* to show what danger lurked in trusting "Fortunes Mutabylitie."[120] He realized that life provided only one temporal end: "For death in a moment consumeth all to dust."[121] Time was the handmaiden of decay, fortune, and death.

Still there was no acceptance of change as natural. Time and decay taught that the true ends to life were not found in the material world they ruled so capriciously, but in the eternal world of God. The social order man maintained was not an end in itself but rather a reflection of eternal order. In the "Induction" to the *Mirror for Magistrates* Thomas Sackville came to grips with mutability, at least on a temporal level:

> It taught me wel all earthly things be borne
> To dye the death, For nought long time may last.
> The sommers beauty yeeldes to winters blast.[122]

Christopher Marlowe may have been the lone Tudor writer who examined the possibility that death and time may claim final mastery over man's life. But throughout the century the reality of change and disorder in man and in society constantly challenged the cosmic vision of a divine and immutable order. Nevertheless, the idea of order remained the conceptual cornerstone of Tudor historical psychology and, indeed, appeared most vital during the last decades of the century when this challenge was most intense. In a significant way, I believe, this helps explain both the peculiar overall political stability of the Tudor years and the positively directed creativity of the Elizabethan years.

The idea of order as the source of an all-encompassing sixteenth-century English cosmic vision, however, enjoyed a somewhat ambivalent success. Those who were its greatest defenders and propagators successfully confronted some anxiety-producing situations, but they ignored many others. New forces such as monarchical absolutism, statute law, and parliamentary sovereignty were too young to require new explanations yet; but their presence was apparent, nevertheless, and the attempt either to ignore them or to incorporate them into the traditional social ordering was merely defensive action. Therein lay the difference between the idea of order as it was described in Ulysses' speech, and as it was described over and over again during the first half of the century. By the last decades of the century the gap between the conceptual idea of order, which was supposed to give meaning to social reality, and social reality was widening. Reiteration of the traditional commonplaces was an attempt to revitalize an old security because no security could yet be expected from the new unrefined values. The attempt continued to use order as it had been used earlier in the century, as a positive conceptualization that defined values and relationships and maintained personal and social security.

Such order and security were definitely precarious. As absolute power spread throughout Europe in the sixteenth century, English theorists alone "produced no theory of the sovereignty of the king, either *solus* or in council or parliament."[123] The growth of parliamentary power and statute law similarly went unheeded by theorists who continued to reiterate fifteenth-century concepts about mixed government and natural law, or based all social relationships upon the divine concepts of authority and obedience. Consequently, in the sixteenth century the conceptual paradigm did not change along with the changes in society.

Political changes indicated changes in personal psychology, too; these changes were not easily incorporated into the idea of order. The Elizabethan years witnessed an increased self-consciousness that was often projected at the group level. Parliament men, merchants, and artists began to experience fulfillment and security in their occupations as occupations. The same feeling motivated behavior on an individual level. The disparity between such feelings and socially accepted values caused a great deal of frustration. Men such as Essex and Bacon and tragedies such as *Edward II* and *Hamlet* bear witness to this fact.

Early in the century Sir Thomas Elyot warned that concepts and ideals isolated from public action bore emptiness. "Know that the name of a sovereign or ruler without actual governance," he said, "is but a shadow, that governance standeth not by words only, but principally by act and example."[124] He had focused, here, on the irony that characterized the Tudor idea of order and, correspondingly, Tudor society. The Tudor English were an assertive, active people; yet the conceptual vision that comforted them failed to recognize such activity as worthwhile. The gap between conceptualization and social reality widened noticeably during Elizabeth's reign. Ulysses' inability to define order in accordance with his own existential exercising of it exemplifies this. In his dedicatory remarks to Walsingham, Ægremont Ratcliffe assured the secretary that "no one may or can leave his owne, to take to him his fellowes office, or charge."[125] This is the ethical imperative of Tudor order theory. Yet in the same dedication Ratcliffe lamented that

> no man is contented with his own lotte: but everie one ledde . . . to undertake things that be contrarie to his naturall instinct . . . whoever saw so many discontented persons: so many yrked with their own degree . . . and such a number desirous, greedie of change, & novelties?[126]

Order theory had become conspicuously coercive. Its definition of social reality clearly ignored changing public and private expressions. The Tudor idea of order survived the sixteenth century because it became incorporated into Elizabethan conservatism; and like Elizabethan conservatism it served to mellow and to refine the potentially radical sensibility of a remarkably creative generation.

Yet within forty years after Elizabeth's death, English conservatism, rather than restraining change, had become a source of radical and innovative theory and action. The history of the reigns of James I and of Charles I has been well recorded, narrated, and interpreted. Social, political, economic, constitutional, intellectual, and religious causes for the Civil War abound. The seeds of restlessness sown during Elizabeth's years and nourished in the following decades reaped a radical harvest during the 1640s and 1650s. W. R. D. Jones' assurance that "by the mid seventeenth century . . . the organic and hierarchical interpretation of society was fading" must surely be considered a reserved yet, nonetheless, an accurate appraisal of the changing conceptions of society.[127]

Without this "organic and hierarchical interpretation of society" the traditional Tudor idea of order had become foundationless. This traditional world view with its sense of an unchanging political order, as we have seen, "precluded any sort of independent political aspiration or initiative."[128] In the old Christian world order "politics ought never to be the concern of private men,"[129] yet that is exactly what politics were becoming during the first half of the seventeenth century.

The evermore conspicuous reality of this political and, as it were, psychological, truth made the redefinition of social meaning and the articulation of a fresh theoretical framework that would facilitate positive and creative consequences essential. Indeed, the idea of order itself began to undergo the most radical metamorphosis. Order was coming to be understood not as natural, but as artificial, created by man, and manifestly political and social. Whereas traditionally order was founded upon the belief that rest was natural, now order theory derived from more empirical political and self-consciousness which allowed that flux and motion were natural; order must be designed to restrain what appeared ubiquitous. So as Tudor order theory had eventually evolved into a negative restraining conceptualization, disregarding political and psychological change, the idea of order developing in the first half of the seventeenth century became a positive restraining conceptualization, based upon political and psychological reality and upon a "new consciousness of politics as a matter of individual skill and calculation."[130]

The result of such individual activity was the articulated sovereign state, the new source of order and consequently the defined locus for personal and social meaning. There is no more striking expression of this new appreciation of individual-social relationship and of the individual's responsibility for defining order and purpose for himself and society than the opening sentences of Hobbes' introduction to *Leviathan*:

> For by art is created that great LEVIATHAN called a COMMONWEALTH, or STATE, in Latin CIVITAS, which is but an artificial man; though of greater stature and strength than the natural, for whose protection and defence it was intended; and in which the *sovereignty* is an artificial *soul*, as giving life and motion to the whole body; the *magistrates*, and other *officers* of judicature and execution, artificial

> *joints; reward* and *punishment*, by which fastened to the seat of sovereignty every joint and member is moved to perform his duty, are the *nerves*, that do the same in the body natural; the *wealth* and *riches* of all the particular members, are the *strength; salus populi*, the *people's safety*, its busines. . . . Lastly, the pacts and covenants, by which the parts of this body politic were at first made, set together, and united, resemble that *fiat*, or the *let us make man*, pronounced by God in the creation.[131]

While Hobbes expanded the trope of organic metaphor so popular in sixteenth-century order theory, he radicalized it, too. The Leviathan resembled man structurally; it could be organically described. But unlike organic resemblances suggested by Tudor order theorists to explain the divine purpose inherent in every activity, Hobbes' metaphor compared the created state to created man and thus man the creator to God the creator. The Leviathan was contracted to do man's will and to fulfill man's purposes and needs; and man fashioned and willed his own order, his own efficient means to peace and satisfaction, his own meaning for activity and existence, and consequently, by extended metaphorical suggestion, his own means of salvation.

Such political theory, describing as it did an entirely fresh idea of order, existentially reduced the gap between conceptualization and social reality that had been widening as long as the traditional idea of order, and all the social and personal values it nourished, was continually, if only rhetorically, articulated. Peace, purpose, salvation, and order now derived from man's individual will. As Walzer notes, as reality and theory merged, "order became a matter of power and power a matter of will, force and calculation."[132]

That order and power began to assume the same characteristics in the minds of politicians and political theorists in the 1640s cannot be considered one of the great riddles of intellectual history. Yet the developing consciousness of a new political reality and of a substantially new psychological identity during the early years of the seventeenth century accounted for this conflation of order and power just as surely as the social crises that preceded the Civil War, and the political and social crises that accompanied it, did.[133] Order overcame the reality of chaos and war, or as theorists began to argue, in one way or another, order overcame the natural tendencies of men or the naturally corrupted tendencies of men. And in such pleading, power, the means to order, became legitimated. Order, then, assumed two qualities that essentially had been the efficient causes for its redefinition: power and repression. Order was the power to maintain social equilibrium and to deny corruption, but it was also the negation of natural instincts and natural liberty.

Hobbes early argued that "the estate of men in . . . natural liberty is the estate of war."[134] Each man, in essence, could do anything to attain his will in nature. Yet because all men had this natural right, no man could really attain his will. There could be no purposeful exercising of one's right if each other one was exercising a similar right. Hobbes argued, then, that peace was really *not war:* it was a negation, essentially a self-denial of natural liberty.[135]

During the 1640s political theorists besides Hobbes naturally concerned themselves with the general relationship between power and order, and with the tangential issues of political legitimacy and the political nature of man. None concluded as clearly articulated a political philosophy of secular sovereignty as did Hobbes, but the questions they posed and the ambivalence with which they confronted the issues reveals much about the developed political consciousness and, consequently, about the new idea of order that characterized the mid-seventeenth century.

Commonly, theorists used secular arguments when discussing the purpose of society, and, in a Hobbesian vein, linked governmental legitimacy to the political nature of man. Man's baseness, a result either of his nature or of his corrupted nature, necessitated a powerful government capable of insuring order.[136]

Anthony Ascham, who published a spate of pamphlets in 1648, took a rather conservative and refined approach to the issue of power and order.[137] The highest aims of society, he argued, were the maintenance of "public peace and quietness" and "the preservation of the commonwealth from destruction."[138] Since the chief end of government was "security and protection," the failure to submit to authority, he claimed in *The Bounds and Bonds of Public Obedience*, resolved that "the commonwealth were dead, and each man were left in his naturals, to subsist of himself and to cast how he could in such a state of war, defend himself from all the rest of the world, every man in this state having an equal right to everything."[139] In other pamphlets Ascham suggested that governments were designed "by man in his public necessities," to preserve order,[140] and that any newly formed government could be considered lawful as long as it supplied "the same law and equity which the excluded magistrates ought to have done, if they had succeeded."[141]

Political circumstances during the 1640s, of course, dictated theoretical cogitations. Ascham's moderating voice, even in support of strictly radical claims, did not represent all opinions. Acceptance of revolutionary governments often necessitated a more radical consciousness; yet rarely was such consciousness unambiguously enunciated. Francis Rous advocated in *The Lawfulness of Obeying the Present Government*, in 1649, that "though the change of a government were believed not to be lawful, yet it may lawfully be obeyed."[142] Such postulation was based, on one hand, upon an uneasy conciliation between an unsteady sense of practical sovereignty and the acceptance of power as order, and, on the other hand, upon a more profoundly traditional religious rationalization that reiterated the Pauline injunction "to obey the powers that be as ordained of God." As Rous put it:

> when a question is made, whom we should obey? it must not be looked at what he is that exerciseth the power, or in what manner he doth dispense it, but only if he have power. For if any man do excel in power, it is now out of doubt that he received that power of God. Wherefore without all exception thou must yield thyself up to him and heartily obey him.[143]

A far more radical and more intriguing polemicist was Henry Parker. Even before the outbreak of hostilities Parker was preparing theoretical justifications for what eventually was to come. In a pamphlet published in 1641 Parker wrote, "Nothing is more desirable to man, or more adequate to the aymes of intelligent creatures than power."[144] Elsewhere Parker denied that law or royal authority emanated from God; rather man constituted them from "pactions and agreements of such and such politique corporations."[145] Man was inherently powerful. By common consent his power might be "derived" into other hands. Man was the "free and voluntary Author," and law the instrument of the transference of power. God confirmed such social activity.[146]

Even a radical theorist needed to concern himself with the relationship between "Absolute Power" and "Order in Nature." For Parker power was expressed experientially. Evidence of power justified the power and the source of power. By this testing method Parker argued that, evidently, priests had no such "power or sword," and by the same logic he concluded in 1643 that Parliament was the legitimate power because Parliament exercised power.[147]

Parker's rhetoric often revealed an ambiguity about such blatant "power" theory. He acknowledged both the need for "Absolute Power" and for an "inherent" source of order for this power. According to "internal operations of the Deitie we ought," he argued "to ascribe prioritie of Order to infinite power."[148] In this manner he sought to justify his belief that social order could result only from the exercising of a supreme and absolute power.

> Power then there must be, and that power must be somewhere supreame that it may command all good, and punish all evill, or else it is insufficient, and if all, then in religious as well as in civill cases, for Supremacie may be severally exercised, but the right of it cannot be severally enjoyed.[149]

While apparently advocating secular sovereignty, the suggested derivation of power from God's infinite attributes tends to compromise the thrust of these remarks. Rather than completely breaking with old conceptions of order and offering new postulates Parker often wrote as if he were still subtly influenced by traditional explanations.[150]

In *Jus Populi*, published in 1644, Parker differentiated between what can be referred to as essential and existential order. Without sin, he argued, "natural order" alone would prevail, but since sin prevailed a suitably defined order which facilitated governmental functioning must be created.

> Order is of a sublime and celestiall extraction, such as nature in its greatest purity did own; but subjection, or rather servile subjection, such as attends human policy amongst us, derives not itself from Nature, unlesse we mean corrupted nature.[151]

This meant that there was a difference between pure essential order and the practical order that confronted sin and corruption. The latter was, itself, an expression of corrupt nature; it was human in derivation and definition and was instrumented through government and law. Necessarily, then, such order was prone to the characteristics of corrupt man. To enjoy the benefits of order man must accept this "burden" that order entailed. So while order emanated from God conceptually, jurisdiction and government emanated from the people.

Hobbes' Leviathan was founded upon the political nature of man and was sanctioned by man's psychological needs. Parker's actively created government, on the other hand, was necessitated by man's corrupted nature and was sanctioned by God's passive permission.[152] God permitted man to choose and to define his own order and his own form of government.[153] Parker differentiated between the "idea" of God as the author of rule and power in nature, the author of "primitive order," and the immediate authority of constitutions.[154] By focusing attention upon "practical order" and upon power and "immediate authority," Parker helped establish new parameters for the ordered, secular, constitutional state.

Just as the Tudor idea of order provided a firm yet resilient foundation for sixteenth-century ideas of society and state, the evolving consciousness of practical and efficient order in the 1640s supported the new conceptions of society and state that were enunciated during the middle years of the seventeenth century. Parker knew that, since man was now responsible for choosing and forming his government, order depended upon man's capacity to create balance and equilibrium in society. Parker, like Hobbes, understood the reciprocal relationship between order and creativity, though unlike Hobbes he did not base his theory of state upon the political and psychological nature of man. Politics and political theory were now concerned only with designing a means to orient peacefully man's power and authority. Parker believed this was possible. In 1642 he claimed that there was now "an Art and peaceable Order for Publique Assemblies, whereby the people may assume its own power to do itselfe right without disturbance to it self, or injury to Princes."[155] Parker here expressed quite clearly the new relationship between "Order" and "Art." Society must be ordered by man; order had to be created; it was not natural, and, consequently, the state, too, was unnatural. It did not reflect any divine purpose or divine order but contractually represented social purpose and articulated social order.

The state began to acquire the attributes both of manifest order and of the source of order. Although many different types of states were described and/or fashioned in the middle years of the seventeenth century, during the first half of the century theorists described a set of general underlying presumptions that radically altered traditional conceptions about the commonwealth, about social purpose, about authority and law, and about the state itself.

Fundamental to this entire reconceptualization of the idea of society was the belief that the commonwealth, as was order, was a human creation. In *De Cive*, Hobbes argued this point politically and epistemologically. Men "are born unapt for society." Without political organization they were merely a multitude of individuals who followed natural instincts and who conformed to no sense or order. Man had no natural bent for community, but opted for society, and contractually articulated social organization when the frustration inherent in following nature became unbearable.[156] By so arguing, "Hobbes' political philosophy . . . expands historical consciousness by embedding the individual in a temporal world and by making him responsible for, as well as tied to, that world."[157]

Commonwealths were demonstrable rather than conjectural as were the phenomena of natural philosophy, said Hobbes. The motions that caused natural phenomena were external to man, but the motions that caused (created) commonwealths were internal motions of men's minds. Men created commonwealths, and introspection led to an objectively certain understanding of these human creations.[158] Man could not learn about politics and order by looking outside of himself; he must turn his attention on himself.

Therefore, Hobbes maintained that no pure, or ideal, form of polity existed; the belief in a natural, divinely ordained, hierarchically structured society must necessarily, then, be fallacious. This traditional conception of the commonwealth he claimed did not derive from any contemporary analysis of politics or society, but was, contrarily, left over from Greek and Roman theory that had merely limned contemporary politics.[159]

Even the philosophically conservative Sir Walter Raleigh advised in his *Maxims of State* that all types of states are just and lawful.[160] He defined state as a "Frame" or the "Set Order" of a commonwealth. (It should be noted that "Frame," a term used often by political theorists during these years, should be understood here to connote definition, boundary, and restraint, as in picture frame, as well as to denote, in its verbal form, creating, shaping, or constructing, as in framing a constitution.[161] Such a broad definition expressed the inherent ambiguity that natural liberty and man's power and creativity caused the new political theorists.) And Raleigh argued that the purpose of the commonwealth was the public good. To reach this end, man created the "Art of Government" or "Policy."[162]

Theorists had long argued that the commonwealth facilitated the common good, but whereas traditionally the common good was understood as the social expression of God's purpose, now the public, or common, good was appreciated as an end in itself. The state functioned well when it accomplished this end. Parker described a "perfect, or intire state," as one "wherein the Governor executes all things in order to the Common good"; and the representatives of the state were really the representatives of the people because "he also is a perfect, and intire Governor, which bends all his actions to that purpose."[163] Parker defined the state by its practical purposiveness rather than by any external and original cause. Such a theoretical stance depended upon an understanding that order was the efficient, implemented cause of a final end rather than the social reflection of a first cause.

Civil order, rather than nature, determined all the personal and the political relations that assured the goal of social, common good. It ordered the relationships between man and woman, and father and children, and even though there was a natural relationship between, say, father and son, acknowledged Parker, it always remained subservient to the relationship defined by authority of the civil order.[164] Likewise the civil order conditioned that the "Prerogative natural of all Kings" was subservient to the "Prerogative legall of the Kings of England," and that this latter prerogative was compatible with, but subordinate to, "popular liberty."[165] Notwithstanding the propagandistic intention of Parker's political comments, he justified his arguments by recourse to the final cause of government. Consequently, Parker depicted the relationship between prerogative and liberty in terms of a reciprocating causal function between means and ends. The liberty of subjects celebrated the king and strengthened the prerogative, while "the Kings prerogative hath only for its ends to maintain the peoples liberty."[166]

Such a "balanced" social organization was conducive to conflicts about power and authority. Though many political theorists had come to identify order with power they had not necessarily been able to describe the most efficient locus of authority that would fulfill the common good. The authority of Parliament versus the authority of the king, and the question of sovereignty, in general, were real issues in politics as well as in political philosophy in the first half of the seventeenth century.

Parker identified a major source of the problem. Since man created social organization to obtain the common good "even that power by which he [king] is made capable of protecting [the people], issues solely from the adherence, consent, and unity of the people."[167] For Parker, final authority, or sovereignty, resided with the body that represented fully the common good, the final goal of ordered, social organization. The king does not retain such authority because he "does not represent the people," and is, consequently, inferior to the people.[168] The Parliament, unlike the king, is not merely a means that endeavors to do the people's purpose, but "is indeed *nothing* else, *but* the very people itselfe *artificially* congregated, or *reduced* by an orderly election and *representation*, into such a Senate, or proportionable body."[169] The Parliament, then, was a willfully articulated body substitute for the "People." It was the organized body of the people that directed their lives by their own consent. Body representation did not refer here to correspondence as it did in traditional Tudor tropology. The Parliament was not an organic resemblance (head and body) of the divine order. Rather, "proportionable" representation was the substitution of an organized body for a nonorganized body, or perhaps, more synecdochically, it was the organization of one body by designating one organ as dominant. In any case the concept illustrated a profoundly enunciated political consciousness. The Parliament was the people self-consciously organizing itself into, and representing itself as, a repressive agent bent upon ordering common good at the expense of nondirectional natural pursuits.[170] Such representation solidified the relationship between order and authority.

Hobbes constantly reiterated this theme: the concepts of sovereignty and representation were cornerstones for his entire understanding of order and the state. He commonly maintained that "concord amongst men is artificial, and by way of covenant."[171] The covenant was the agreed-upon organization articulated between *sovereign* and *subject*, as Hobbes finally described it in the nineteenth chapter of his early work on the *Elements of Law*.[172] In the *Leviathan* Hobbes defined the sovereign as the body that controlled the power in the commonwealth. Unlike Parker, Hobbes refused to designate any particular body as the one in which this power resided, but he did argue that a covenanted, instituted commonwealth was the *representative* of the multitudinous will.[173] Hobbes, too, radically changed the organic sense of representation that characterized Tudor conceptualizations, and posited an extreme nominalist view of sovereign representation. The sovereign represented the soul of the community;[174] social and political organization ordered and governed by force of will rather than by reason.

Because he enabled man to live outside the precovenanted condition of chaos, the sovereign assumed responsibility for each man's end as it became the public good. Hobbes' sovereign, then, was the authority that conciliated means and ends into one ordered society. The sovereign could not be punished because, in reality, he was merely acting for the subject. Hobbes developed a convenient sense of an extended burden of action that eased the responsibility inherently experienced in accepting authorship of one's own destiny.

Conclusions about sovereignty and authority required reevaluation of the traditional understanding of law. If man chose his own form of government and created his own social ends then it followed that he "made" laws to guide his political order. The history of the development of Parliament from a juridical into a legislative body evidenced that, constitutionally, this was indeed, what happened. But in order to accept that man created law it was necessary to overcome the traditional view that law was eternal and that civil law merely reflected and made known eternal law. An acceptance of the created quality in law facilitated the political consciousness necessary to define personal and social purpose around the secular state.

Early in the seventeenth century Raleigh wrestled with conflicting views of law advocated by Royalist sympathizers and common lawyers. Unable, or perhaps unwilling, to choose one position over the other, Raleigh nevertheless refused in his discussions to elaborate the opinion that civil laws reflected eternal laws. "All Laws," he wrote, "whereby Commonwealths are governed, were either made by some one excellent Man, and at an Instant; or else they were ordained at sundry Times according to such Accidents as befel."[175]

Hobbes, too, maintained that civil laws were made. Neither law nor justice existed before men relegated power to a sovereign. "Justice and injustice . . . are qualities that related to men in society not in solitude."[176] Man's natural passions, once repressed through political organization, became the foundation for a system of justice and the concept of natural law.

In the *Elements of Law* Hobbes maintained that "all laws are declarations of the mind, concerning some action future to be done, or omitted."[177] Epistemologically, this stated that law resulted from the motions of the mind, and was only creatable and knowable by man. But since Hobbes argued that law resulted from the transference of power from natural man to social man represented by a sovereign, he later modified this statement so that it referred to "Laws of Nature," which were "nothing else but certain conclusions, understood by reason, of things to be done and omitted."[178] By laws of nature Hobbes referred here not to fixed external rules, but to fixed internally motivated concepts of behavior such as the concepts of appetite and aversion that he discussed in *Leviathan*.[179] Consequently Hobbes maintained the idea of eternal, natural laws, but rid it of any external, nonhuman origin. And even though these laws of nature were "immutable and eternal" they were not necessarily limiting and defining because "actions may be so diversified by circumstances and the civil law, that what is done with equity one time is guilty of iniquity at another."[180] Since both the law of nature and civil law shared as a prerequisite the establishment of the commonwealth, Hobbes declared that they "contain each other, and are of equal extent."[181] Once the state was covenanted civil law was written to provide the commonwealth with the civil order necessary to maintain the public good.

As he determined the nature of political reality from the political and psychological nature of man, so too, Hobbes understood that "law" derived its existence from this same source. In refutation of Coke and others who argued that law could never contravene reason Hobbes ascertained that these theorists never delineated what reason was. For Hobbes, though, the only reason was "this of our artifical man the commonwealth, and his command, that maketh law."[182]

New concepts and definitions of the state, of the commonwealth, and of law were accompanied by a wholesale reconsideration of human psychology and of personal and social values and ethics. What motivated human behavior; how was it directed? If law and order were socially directed could "right reason" still activate them? And what was reason now that man began to take responsibility for his own ends? How should behavior be evaluated in this practically oriented world? Indeed, against what constant frame of reference could anything be evaluated if motion was natural and if law and order were human creations contrived to fulfill the human goal of peace and public weal?

For Hobbes the concept "right reason," used in its traditional sense, suggested equivocation. Generally, Hobbes understood reason to be the ability to reason well.[183] He abandoned the Tudor concept of "right reason" as a divine virtue externally directing human behavior toward rational goals. Borrowing from the recent studies of William Harvey, Hobbes determined that the heart was the central and controlling organ in the body. The passions, then, motivated human behavior since the heart determined perceptions, feelings, and desires. Man desired what facilitated the circulation of the blood. This materialist viewpoint (Hobbes believed that bodily activities and psychological phenomena operated the same way—indeed were the same) which argued that the mind was "body-dependent" substantiated Hobbes' theory that all men were uniformly and internally motivated.[184] The negative sense of passions that Tudor commonplaces advanced and even the more balanced psychology advanced in the later Elizabethan years were inimical to this view.

Passions and reason were not antithetical, but on the contrary, reason derived from passion and served passion by directing behavior. Generally Hobbes argued that men's natural passions were dualistic; one was for gain and appetite, the other for self-preservation, or fear of death. This second passion dominated the first and naturally "reasoned" a desire for peace. As Oakeshott understood, reason was not the "arbitrary imposition upon the passionate nature of man; indeed it [was] generated by the passion of fear itself."[185]

Men obeyed rules and laws of society not because reason commanded it, but because security and peace could be constructed by obeying rules and that was reasonable. As Hobbes declared in *De Cive*, "to seek peace . . . is the dictate of right reason," which is not "an infallable faculty."[186] Reason was merely the act of reasoning properly about one's actions, and if exercised it would direct or project one's will and desire through peace toward the goal of self-felicity. In civil society the supreme law was right reason, Hobbes argued.[187] Outside of civil society right reason could only be gleaned by comparing one's own reasoning and behavior with another's. Finally, then, right reason revealed those laws of nature which, by repressing those natural instincts that motivated activity, directed behavior toward satisfying the passion for self-preservation.

The state, directed by reason, restrained and satisfied the passions simultaneously. Hobbes devised a healthy way to reduce the burden of responsibility that accompanied the charge of fashioning one's own security and salvation. In the introduction to *De Cive* he compared men in nature to children. Their nature was not evil; it was merely unorganized and nondirected. By this he meant that their nature was not positively repressed. Neither men in nature nor children have "the free use of reason."[188]

In the *Leviathan* Hobbes discussed another aspect of reason. "Reason, or *reckoning*, that is adding and subtracting, of the consequences of general names agreed upon for the *marking* and *signifying* of thoughts,"[189] enabled man to evaluate things; "there being nothing in the world universal but names, for the things named are everyone of them individual and singular."[190] Men deceived themselves "by taking the universal, or general appelation, for the thing it signifieth."[191] Hobbes attempted to establish an ethical means for evaluation around this extreme nominalist position by combining it with secular, temporal, and utilitarian interests. Reason was "attained by industry" and characterized by a type of deductive Galilean methodology, which resulted in *science*, the "knowledge of all the consequences of names appertaining to the subject in hand."[192]

Since there was nothing universal against which to evaluate anything, there was nothing infinite either. "Whatever we imagine is *finite*."[193] Hobbes maintained that the infinite connoted only one's inability to conceive boundaries and one's incapacity to evaluate the thing in itself. Just as he argued that justice and injustice were relative and finite evaluations determined by specific civil circumstances, so Hobbes determined that true and false and good and evil were specific attributes "to be taken from the nature of the objects themselves."[194]

Hobbes confronted the same issues that Marlowe and Shakespeare dramatized a generation earlier. Tamburlaine attempted to name himself by his finite behavior: "I am my deeds." And the Trojans and Greeks, in *Troilus and Cressida*, sought value and worth in Helen and in battle, rather than in some general conception. Hobbes secured such finite and existential ethics by methodologically marrying them to a secular eschatology that he founded upon man's natural proclivity to individual self-satisfaction. Proper evaluation and knowledge of things, then, facilitated social and personal felicity. As Hobbes affirmed, "*reason* is the *pace*; increase of *science* the *way*; and the benefit of mankind, the *end*."[195]

Such a radical and methodological means of evaluation meant that individual and social worth and importance were judged against new standards and by new processes. Certainly the traditionally ordered social structure could not be justified by these new views. Nor could such external concepts as "understanding," "virtue," and "right reason" determine individual worth.

Neither value nor worth was absolute, "but a thing dependent on the need and judgment of another."[196] A man's worth was his price. If he was powerful he was honored, and consequently, he was honorable. He could barter power for esteem. Man's attributes acquired value, Hobbes suggested, only in social relations.[197]

Riches and good birth were honorable because they indicated power and facilitated advancement toward desired ends; and similarly, poverty was dishonorable because it signified weakness. "Honor consisteth only in the opinion of power."[198] Titles, then, were honorable because they signified "the value set upon them by the sovereign power of the commonwealth,"[199] not because they marked the predetermined social order in any divine chain of hierarchic status.

Prevalent concepts such as "virtue" were meaningless now. Hobbes' ethics radically altered the traditional Tudor commonplaces that determined meaning and value. "What is," not "what should be," formed the basis of social meaning and personal evaluation. Practicality defined worth. Hobbes established a systematic ethics based upon optimally utilizing man's natural desires and tendencies. In doing this he tied the security and order of the public realm to the aspirations of each individual and provided the individual with the rationalization needed to continue to exercise his capacities.

In this way Hobbes fashioned a positive and internally created relationship between man and society to replace the traditional conception of man's predefined place in society that had been the source of much recent frustration. Hobbes argued that all men were equal in nature; only civil society through civil law determined inequality.[200] He confirmed no known *order* for degrees among men. Reasons might exist for inequality but society must realize that a natural law existed that "every man acknowledge [the] other for his equal."[201]

The dictum "read thyself" encouraged social harmony, Hobbes declared, because it led the individual to understand his place in society by exposing "the similitude of the thoughts and passions of one man, to the thoughts and passions of another."[202] Men were actively similar but passively dissimilar. They all exercised hope, fear, and desire, but objectified these passions differently. Consequently, the state that men contractually articulated necessarily represented and directed their passions.

In concluding his discussion of the "Tudor Commonwealth" W. R. D. Jones described the English state in the mid-seventeenth century:

> The State may be envisaged either as an entity supreme in itself or as an agent for the preservation of the individual rights of its members. It may, in either case, be appealed to on grounds of power and wealth, criteria whose validity would not have been accepted within a medieval religious context, rather than on those of equity and morality.[203]

What Jones only hinted at here was the development of the state as the source that nourished the idea of order and that defined all the concomitant personal and social values identified with order. As the state's function expanded in this manner, the individual's consciousness of his responsibility for his own destiny developed too. The smoothly functioning institutionalized secular state facilitated man's desires while at the same time it relieved the individual of the *direct* consequences of this burden.

In *De Cive* Hobbes concluded that "one name alone . . . doth signify the nature of God, that is, existent."[204] Man could name God but he could not know him ontologically. He could not assign any ideas, qualities, actions, or passive faculties to God since all that had significance for man was finite and temporal and could be referred only to specific things with specific attributes. Consequently, nothing human could be explained in reference to God.

Henry Parker evaluated human behavior in reference to the final end of goodness for mankind. Anything that furthered this end Parker declared to be natural and good. He made "necessitie" the foundation for "Policie" and argued that man needed "more particular constitutions" to bind him than the simple sense of "Order" and divine and natural law.[205] Although he posited no theory of natural motion, as Hobbes did, Parker, too, deduced the necessity of evaluating human behavior within the confines of the secular, temporal sphere by uniting his keen sense of political reality with the received idea of the Fall. In worldly affairs, such as politics and government, Parker warned, man "ought to bee very tender how we seek to reconcile that to Gods law, which we cannot reconcile to man's equity; or how we make God the author of that constitution which man reaps inconvenience from."[206] Though Parker did not psychologically and philosophically separate man from God's world, he did place the responsibility for man's "worldly affairs" heavily upon man's shoulders.

The most conspicuous consequence of substituting the secular state for the "divine cosmos" as the source of order, security, and meaning in the world was a heightened consciousness of personal mortality. As Hobbes noted in *De Cive*, temporal death was to be feared as greatly as eternal death.[207] Man's awareness of motion and flux as natural, and of order, security, and immutability as contrived, led him self-consciously to identify purpose and meaning in the temporal world. Man lived for life not life after death. "I put for a general inclination of all mankind, a perpetual and restless desire after power, that ceaseth only in death," said Hobbes.[208] Man aggrandized in order to secure what was not securable.

The Tudor idea of order defined a static society that conceptually assured peace, security, and salvation. The idea of order that developed during the first half of the seventeenth century promised security and peace, but the psychological need that motivated it neither allowed a static society nor understood that peace and salvation were reconcilable. Though men must live in peace, Hobbes believed that the "felicity of this life consisteth not in the repose of a mind satisfied. . . . Felicity is a continual progress of the desire, from one object to another."[209] Mortality defined the need for order and described the realm of salvation.

CHAPTER 2

The Dramatic Microcosm: From Tamburlaine to Prospero, The Struggle with Self-Definition

TAMBURLAINE: I am a lord, for so my deeds shall prove.
Christopher Marlowe,
Tamburlaine the Great
I *TAM* I.ii.

MACBETH: What is't you do?
ALL (witches): A deed without a name.
Shakespeare, *Macbeth*
IV. i. 48–49.

PROSPERO: . . . we are such stuff
As dreams are made on; and our little life
Is rounded with a sleep. . . .
Shakespeare, *The Tempest*
IV. i. 156–158.

Lord Whitehead's advice that "it is to literature that we must look, particularly in its more concrete forms, if we hope to discover the inward thoughts of a generation,"[1] rings profoundly true when the generation under surveillance is Shakespeare's: a generation which became increasingly and articulately self-conscious.[2] Yet so many who do indeed "take their notion of the Elizabethan age principally from the drama," ignore this self-consciousness.[3] Shakespeare himself is often acknowledged to be a conservative and a representative Elizabethan, ruled as was his age "by a general conception of order . . . so taken for granted, so much part of the collective mind of the people."[4] Indeed, so unself-conscious are the Elizabethans pictured to have been that Tillyard maintains they shared "a mass of basic assumptions about the world, which they never disputed"; and these assumptions described a world picture that was not only dogmatically traditional, but that was, indeed, "a simplified version of a much more complicated medieval picture."[5]

Such a myopic interpretation of Shakespeare and of Elizabethan culture results in part from treating textual reality as representing contextual (cultural) meaning. Rather, I believe, we can examine these texts as processes that participate in the production of meaning. No more than did Hooker or Bacon did Shakespeare convey a feeling only for a fixed and traditional world picture. His theater, I suggest, dramatizes a shift from consciousness to self-consciousness, a shift from meaning-in-being to meaning-in-becoming. As one critic argued, his drama

> can be seen as part of the process by which our culture has moved from absolutist modes of thought towards a historical and psychological view of man. . . . And he wrote at a moment when the educated part of society was modifying a ceremonial, ritualistic conception of human life to create a historical, psychological conception.[6]

It intimates a "historical" meaning for human experience which accepts the definitive nature of man-in-time and articulates a relationship between identity and social order that is defined by the individual's consciousness of, and responsibility for, self. I am not arguing that Shakespeare focused upon a particular theme or that his "canon" *means* this or that. Rather I imagine his theater as discursively mediating a historically creative process of meaning redefinition which suggests the substitution of a self-consciously defined concept of secular order and identity for the received idea of a divine cosmos. Following from this, an interpretation of the process of this theatrical vision as it developed over time will reveal neither the meaning of Shakespeare nor the meaning of any of his plays; rather it should highlight areas of psychological conflict which ultimately characterized the lived process of social and historical change. As Levao noted, "Shakespeare's aim is not subversion but a dramatic exploration of Tudor ideals as expressions of human desire, a part of historical process rather than its authorized explanation."[7]

Before turning to Shakespeare, however, it will be useful to look briefly at an example of traditional Tudor drama, Norton and Sackville's *Gorboduc*, for another description of the commonplaces of society and the idea of order. It will also be of benefit to look at Marlowe's plays for an expression of a spontaneous and unrefined attack against the traditional order. Marlowe's characters experience less self-consciously psychological and historical burdens similar to the ones which Shakespeare's later heroes do. They find it necessary to impose order upon a hostile world,[8] but they are unsure about locating the source of order in themselves. They are "moral and intellectual voyagers who have cast off from the psychic boundaries of the Old World, and only after it is too late do they recognize that they are lost."[9]

Gorboduc is a didactic play whose reserved tone does not fully hide a sense of desperation. A conscious effort throughout the play to reinvoke the traditional social values of order and degree manifestly accentuates an inherent insecurity regarding these same values. The wise counselor, Philander, warns his king, just as the clever Ulysses tells Agammemnon in Shakespeare's *Troilus and Cressida*, about the inevitable consequences that accrue from failing to abide by order and degree:

> Only I mean to show by certain rules,
> Which kind hath graft within the mind of man,
> That nature hath her order and her course
> Which (being broken) doth corrupt the state
> Of minds and things, ev'n in the best of all.[10]

In a dumb show that serves as a thematic prologue to the play, Norton and Sackville outline the accepted commonplaces of the traditional order. "Hereby was signified, that a state knit in unity doth continue strong against all force, but being divided is easily destroyed" (p. 407). Throughout, the necessity of good counsel is reiterated; and the effects of bad counsel are dramatized in the persons of Hermon and Tyndor, the respective "Parasites" advising Ferrex and Porrex. "As befel in the two brethren, Ferrex and Porrex, who refusing the wholesome advice of grave counsellors, credited these young parasites, and brought to themselves death and destruction thereby" (p. 409). Chaos occurs directly "when kings will not consent to grave advice but follow willful will" (p. 452).

The threat of mutability, civil war, and chaos haunts *Gorboduc*. The initial disregard for order, Gorboduc has matched "his younger son with me / In equal pow'r, and in as great degree" (pp. 420–421), Ferrex complains, generates unnatural pride and ambition, and inevitably leads to social disorder. Still, subjects cannot rebel, because only by resisting more change can society resist destruction. Positive, alternative modes of behavior are not suggested here. Meaning is absolute. There is order and there is disorder; and both private and public identity reflect this.[11]

Marlowe's theatrical world is far more self-contradicting and confusing. The basically medieval and Christian social structure exists but it is rather less than fully functional. Marlowe's characters possess individual aggressive energy and the temperament to travel beyond normal ethical boundaries. In fact they ethically govern their own worlds, which at times appear morally empty. Barabas, for example, moves about in a frenetic world from which all love, loyalty, and restraint have apparently gone missing. The end result is rapid unrefined energy.

But Marlowe's theater hyperbolically dramatizes confusion and insecurity, and is, in fact, an ethical demonstration of the burdens of self-definition. His characters desire to accumulate. Barabas covets wealth, Faustus knowledge, Edward Gaveston and Tamburlaine power. Each submits completely to his single purpose, yet for each the goal is merely a means to an end: salvation and meaning in life. In confronting the burden of self-responsibility in a world lacking Christian retributive justice, each character begins to locate ultimate meaning—salvation and damnation—in consciousness, in history, in "temporal, secular, existential conditions."[12]

Tamburlaine celebrates individual private desires. Virtù and will fashion his personality and direct his actions. He definitively steps out of "order" and "degree" first by asserting himself a "greater man" than the Soldan, and then by defining his own exalted position through his actions: "I am a lord, for so my deeds shall prove" (I *TAM* I. i. p. 7). In creating himself, he assumes a full burden from which he can not escape. His self is defined as much by "shall" as by his actions. The tragic potential of such self-definition quickly arises. Tamburlaine forsakes Zenocrate's wishes because he has to honor his own imperatives:

> Z: Yet would you have some pity for my sake,
> Because it is my country and my father's.
> TAMB: Not for the world, Zenocrate, if I have sworn—
> (I *TAM* IV. ii. p. 41)

Marlowe defines what Shakespeare and Bacon refine: the need to make personal submission social and utilitarian in order to escape the tragic consequences of self-definition.

Tamburlaine, after donning Cosroe's crown, the symbol of earthly power and felicity, asserts:

> So; now it is more surer on my head
> Than if the gods had held a parliament
> And all pronounc'd me King of Persia.
> (I *TAM* II. vii. p. 24)

Indeed, Tamburlaine and his men agree that not even God in Heaven enjoys the glories or pleasures that a king on earth does. Pride and ambition circumscribe the secular salvation sought in kingship. As Tamburlaine wishes, "May we become immortal like the gods" (I *TAM* I. ii. p. 11).

Tamburlaine acknowledges no nemesis. But he does not accept the full burden of his own self-responsibility either. By continually justifying himself as God's scourge, he denies authentic self and consequently limits his aspirations to a common, externally defined goal: kingship. And because of this he constantly confronts the evanescent nature of the unrefined identity he claims. Zenocrate recognizes his insecurity and his limitations:

> Ah, Tamburlaine my love, sweet Tamburlaine,
> That fights for sceptres and for slippery crowns,
> Behold the Turk and his great emperess!
> (I *TAM* V. i. p. 55)

Her ties to the old order and her creative love offer Tamburlaine a chance to solidify his identity by emancipating it from the burden of proving actions. As the first part of *Tamburlaine* concludes, he is still obsessed with securing his self in aggressive and continuous action:

> Hell and Elysium swarm with ghosts of men
> That I have sent from sundry foughten fields
> To spread my fame through hell and up to heaven:
> . . . . . .
> And such are objects fit for Tamburlaine,
> Wherein, as in a mirror, may be seen
> His honour, that consists in shedding blood
> When men presume to manage arms with him.
> (I *TAM* V. I. p. 58)

But the final reconciliation scene, in which Tamburlaine pledges peace with the world, respectfully buries the symbols of the old order, the "great Turk and his fair Emperess," and marries Zenocrate, suggests that Tamburlaine may measure rest with action and love with power and thereby refine the energy with which he defines his own identity, and consequently, new social values and a new order.

Yet Zenocrate's death early in the second part forces Tamburlaine to retreat into himself and insures that for him death alone monopolizes the ultimate meaning of life. The entire second part of *Tamburlaine* concerns the consciousness of mortality. And when Tamburlaine dies too, all energy and order expire, and salvation with them.

> But, if he die, your glories are disgrac'd,
> Earth droops, and says that hell in heaven is plac'd!
> (2 *TAM* V. i. p. 114)

The final note resounds despairingly and hollowly. Wherein lies redemption? "Meet heaven and earth, and here let all things end" (2 *TAM* V. iii. p. 119).

No play in Marlowe's canon so purposively questions the meaning of life and the meaning of death as does *Doctor Faustus*. Faustus' search leads him not to final redemption, but rather to an appreciation of the limits of secular possibilities and to an ambivalent awareness that such limits define salvation and damnation as one. Faustus' quest, however, is inhibited by a conspicuous lack of awareness concerning his own creativity and his own limitations, and develops apace as does his self-consciousness.

Like Tamburlaine, the unself-conscious Faustus aspires not beyond but beneath his capabilities. Tamburlaine seeks felicity in an earthly crown. Yet this is but an acceptable example of the old order. It is his desire and his will, his power to name himself and to define his world that mark his true potential. His private desires and actions far outreach the crown he reaches. So too does Faustus seek felicity; but he expects salvation or damnation from traditional sources. His reason and will do not direct him to seek knowledge beyond the limits of the known. He puts his knowledge and will to the task of securing information about otherworldly issues.

The creation of a new world picture requires the destruction of the old one. Faustus' identity, however, emanates from the Christian world view. Yet his energy and his will spontaneously react against it. In struggling to establish temporal, secular values with which to define purpose in the world, Faustus simultaneously begins to reject religious values and to free himself from the power they exert over him. The extended metaphor of appetite which runs throughout the play turns the Christian communion sacrament into a type of secular eucharist as Faustus attempts to become one with the world.[13] He begins to recognize the Hobbesian nature of will and "appetite," but he is yet unprepared to utilize it in any socially purposive way. As he reminds himself:

> The God thou serv'st is thine own appetite,
> Wherein is fixed the love of Beelzebub.
> (II. i. 11–12)

Faustus partakes of his "last supper," then, before he begins to move away from Christ and symbolically remains close to him at the same time. His blood congeals as he becomes a sacrifice to the devil. He is on his way to secularizing salvation through the medium of his own self-sacrifice.[14]

Throughout the play he contemplates mortality and redemption. He tells Valdes, "as resolute am I in this / As thou to live" (I. i. 135–136). But his conscience denies him peace of mind. Faustus' only link to Heaven is the Good

Angel, and by any analysis it is a specious link. The Good Angel's first appearance marks it for all time as illusory, as Faustus' imagination:

> O, Faustus, lay that damned book aside,
> And gaze not on it, lest it tempt thy soul.
> (I. i. 71–72)

This after-the-fact advice bespeaks a troubled conscience, not a holy emissary. At one later point in the play Faustus begs Christ for salvation but receives instead a friendly visit from three fiends of Hell. And soon thereafter, he questions: "Accursed Faustus, where is mercy now?" (V. i. 70).

C. L. Barber locates Faustus' self-consciousness in the poetry. "The hero constantly talks about himself as though from the outside, using his own name as to develop a self-consciousness which aggrandizes his identity."[15] For example: "Settle thy studies, Faustus, and begin" (I. i. 1). Is this truly self-consciousness? Although the poetic structure suggests that Faustus "aggrandizes his identity," the meaning of the lines qualifies this view. After the devil changes shapes upon Faustus' request, Faustus remarks:

> How pliant is this Mephistophilis,
> Full of obedience and humility.
> Such is the force of magic and my spells.
> Now, Faustus, thou art conjuror laureate.
> (I. iii. 29–32)

Faustus addresses Faustus the magician, here. In recognizing magic as the source of his power, he misrepresents and underrates his own definitive self-will. Necromancy is no more the source of Faustus' powers than "scourge" is of Tamburlaine's.

Faustus commands himself: "here try thy brains to get a deity" (I. i. 63). When he decides on magic he becomes excited, "ravished," and cries, "O, this cheers my soul!" (I. i. 150). He desires to "conjure in some lusty grove / And have these joys in full possession" (I. i. 152–153). Later he admits that appetite is his god. It is in this attempt to raise the soul through the powers of the flesh, to locate the spiritual in the temporal, that Faustus charts his course to salvation, damnation, and self-consciousness. A. O. Lovejoy describes the confrontation between the medieval commonplace of outer-worldly identification, and an inner-secular identification based upon variety, creativity, and nature's process.[16] This confrontation, he claims, facilitates the appreciation that "temporal realization of the absolute reaches its highest point in man, a being capable of self-consciousness";[17] in essence, the knowledge that "Man is God wholly manifested."[18]

No matter how much Faustus refers to the Christian world of Heaven and Hell, salvation and damnation, finally his imagination situates them in a very different world: one of social relations. The devils appear before Faustus conjures. It does not require evil actions to summon devils in this world; contemplation will do. The magic is superfluous. As Mephistophilis notes:

> Therefore the shortest cut for conjuring
> Is stoutly to abjure the Trinity
> And pray devoutly to the prince of hell.
> (I. iii. 52–54)

This is a world of metaphysical questions. Who is this "prince of hell?" To whom has Faustus prayed except to his own desires? And doesn't he admit as much?

If Faustus seeks salvation in worldly bliss, and if he conjures the devil by his thoughts, then an affinity exists between Heaven and Hell which transcends the locus of the Christian relationship. Faustus boasts:

> This word "damnation" terrifies not me
> For I confound hell in Elysium.
> (I. iii. 59–60)

He refuses to separate Heaven and Hell. When Mephistophilis explains, "why this is hell, nor am I out of it" (I. iii. 74–75), Faustus scorns him because he is "passionate / For being deprived of the joys of heaven" (I. iii. 82–83). Faustus then admonishes him to "scorn those joys thou never shalt possess" (I. iii. 86). For Faustus they do not exist. Faustus discusses Hell with Mephistophilis, again, a little later, and calls it a fable. Mephistophilis maintains:

> For I tell thee I am damned and now in hell.
> (II. i. 135)

And Faustus quickly replies:

> Nay, and this be hell, I'll willingly be damned.
> (II. i. 136)

Faustus cannot know, no matter how much knowledge he gains, the psychological torments of a hell of one's own creation without a more fully developed self-consciousness. Unlike his mentor, he has not yet experienced the fugacious nature of temporal heaven. This explains Mephistophilis' initial "reluctunce to dwell upon the suffering that Faustus cannot grasp as real."[19]

The poetry of the "Helen" scene is perhaps the most passionate, the most spiritual poetry in the entire Marlovian canon. It is here that, for an ephemeral moment, salvation and damnation, Heaven and Hell, body and soul are indistinguishable, are one.

> Sweet Helen, make me immortal with a kiss
> Her lips suck forth my soul. See where it flies!
> Come, Helen, come give me my soul again.
> Here will I dwell, for heaven is in these lips,
> And all is dross that is not Helena.
> (V. i. 101–105)

Helen transports Faustus beyond mortality, beyond meaninglessness, with her kiss. The kiss is like communion; touching the source of power, it promises immortality.[20] Simultaneously, Faustus loses a heavenly soul and gains an earthly one. Helen embodies his Heaven. Not only does he combine his Heaven with his "Voluptuous" Hell, but he substitutes this secular salvation for traditional redemption. Barber notes the similarity between the poetic movement of the last line just quoted and Mephistophilis' earlier description of Heaven:[21] "All place shall be hell that is not heaven" (II. i. 124).

As Faustus' end draws near, his earthly bliss is as swiftly gone as enjoyed. Mephistophilis notices that

> . . . his *laboring* brain
> *Begets* a world of idle fantasies
> To overreach the devil.
> (V. ii. 12–14; emphases mine)

Can Faustus "conceive" his own end now? To overreach the devil he must accept the full "burden" of his own self-responsibility. He must relegate redemption to the past; he has authored his own ends; he must grasp his own Truth. As the clock strikes twelve, amidst thunder and lightning, Faustus makes his final plea.

> O, it strikes, it strikes! Now body, turn to air
> Or Lucifer will bear thee quick to hell.
> O, soul, be changed to little waterdrops,
> And fall into the ocean, ne'er be found
> . . . . . .
> My God, My God, look not so fierce on me!
> Adders and serpents, let me breathe awhile.
> (V. ii. 180–185)

Does Faustus finally repent? It appears not. Does he appreciate his own powers, accept his own creativity? This, too, seems unlikely. Without doubt, he clings to body and soul. They are not divided; the one is saved in and with the other. Barber equates this scene with the sexual consummation of the "Helen" scene, referring to the flaming Jupiter and hapless Semele, and to Arethusa.[22] Faustus, then, at the end recalls the moment of his earthly salvation and tries to reinvoke its magic, to fix it immutable. In this case, he has transcended the confines of the old order. He confronts the existential void, the transience of one's own created universe. Erich Heller maintains that Faustus' search led him to a truth that was Hell.[23] He dies on the threshold of a new universe which will offer to isolated individuals temporary salvation and permanent psychological damnation.[24]

In *The Jew of Malta* and in *Edward II*, Marlowe most profoundly dramatizes the burdens of this developing sense of isolated selfhood. Barabas constantly acknowledges that he must "look—unto myself" (I. p. 166). No matter what happens in the outside world, he will "make sure for one" (I. p. 167). But his isolation is not totally self-imposed. The governor, Ferneze, singles him out as an individual sacrifice: a Jew sacrificed to save a Christian commonweal.

> No, Jew; we take particularly thine,
> To save the ruin of a multitude:
> And better one want for a common good,
> Than many perish for a private man.
> (I. p. 170)

And Marlowe uses the Jewishness to incorporate the felt suffering of isolation. Barabas celebrates an aloneness that he cannot resist. In a line which recalls the poetic sense of Faustus' temporal salvation, Barabas concludes: "But all are heretics that are not Jews" (II. p. 188). The Christian world in which he must live and from which he is ever alienated makes life Hell.

As an outcast, and specifically as a Jew, Barabas emphasizes the importance of individual deeds and actions. Not willing to have his life judged against any preconceived evaluation of what Jews might have done to Christ, Barabas demands that he be judged on his acts alone.

> But say the tribe that I descended of
> Were all in general cast away for sin.
> Shall I be tried for their transgression?
> The man that dealeth righteously shall live;
> And which of you can charge me otherwise?
> (I. p. 170)

And it is for his acts that he is finally judged. Barabas proudly accepts the responsibility for his death. Intent "To end thy life with resolution" (V. p. 223), he self-consciously details his past behaviors and boasts about what he had planned yet to do. But the acts he owns are all negative. His will and his self-consciousness are still too unrefined to describe a positive, satisfactory self-defined purpose and order for life.

Like Barabas, Edward, too, celebrates his private self because the identity which fashions his public self is too inhibiting and is ultimately nonfulfilling.

> I'll bandy with the barons and the earls,
> And either die or live with Gaveston.
> (p. 229)

He sets his self against the world and tries to secure it in his love for Gaveston. Edward II crosses the bridge from the outer to the inner world of meaning and salvation when he so beautifully answers the haughty Mortimer's question about why he loves a man whom the whole world hates: "Because he loves me more than all the world" (p. 235). The emptiness of his public identity weighs heaviest upon him when Gaveston leaves. He is unable to be a king and unable, too, to be himself without his love: "Thou from this land, I from myself am banish'd" (p. 236). Indeed, Gaveston's death alone spurs him to any concerted action. He cannot invoke any of the magic usually associated with the nature of kingship.

> But what are kings, when regiment is gone
> But perfect shadows in a sunshine day?
> (p. 277)

And without the object of his love he cannot sustain his private subjective self either. Edward imparts nothing of himself into his kingship, but rather, tries to suck some final sustenance from the symbol of his public identity, his crown.

> See, monsters, see! I'll wear my crown again
> What, fear you not the fury of your king?
> (p. 278)

Indeed, it is to his public identity that he turns when confronted by his inevitable death. The private self he has stroked for so long has not prepared him for his end. Even more vehemently than Faustus did at his end, Edward wishes to resist the final authentic burden of his own acts. He begs time to stop so that he may remain alive and England's king. Shortly thereafter, however, under pressure from the bishop of Winchester, Edward gives up his crown and with it his kingship. Left alone, with only this final negative act, Edward finally seeks to escape his inner self, too.

> Come, death, and with thy fingers close my eyes,
> Or, if I live, let me forget myself!
>
> (p. 279)

Marlowe's drama evokes an old world, decaying and nonredemptive, to which one cannot easily look for meaning in life, but to which, still, one has to explain and justify action taken to locate new meaning. There exists "anxious uncertainty or a corrosive skepticism deriving from the awareness that the religious foundations of medieval philosophy and the age-old assumptions of order and degree were being superseded."[25] But the energy expended in reaction to that anxiety is not merely privately dangerous and most often unfulfilling, it exacerbates the weakness of the old order too. Anomalous values do not immediately provide for a new order, new salvation, but they do insure that the old order will never recover from its initial decay. As Harry Levin notes, "While science, capitalism, [and] imperialism were at the beginning of their modern development, Marlovian tragedy was able to project the inordinate courses they would pursue."[26] Subsequently the refined and rationalized energies of a Marlovian self-conscious individualistic ethic would define the shape of social order and meaning.[27] But for Marlowe's heroes, themselves, the time was not yet. The existentialist philosopher Karl Jaspers describes the nature of the tragedy of transition. It is the nature of Marlowe's Elizabethan world.

> The old [order] is justified in asserting itself, for it still functions; it is still alive and proves itself through its rich and elaborate traditional patterns of life, even though the seed of decay has already begun its fatal germination. The new is justified also, but it is not yet protected by an established order and culture. For the time being it still functions in a vacuum. But it is only the hero, the first great figure of the new way of life, whom the old, in a last frantic rally of all its forces, can destroy. Subsequent breakthroughs, now untragic, will succeed.[28]

Shakespeare's idyllic comedies, *A Midsummer Night's Dream* and *As You Like It*, share in common release from the confrontational realities of time creating illusionary, dreamlike worlds which starkly and conspicuously contravene the world of confused social reality.[29]

> THESEUS: Merry and tragical? Tedious and brief:
> That is, hot ice and wondrous strange snow—
> How shall we find the concord of this discord?
> PHILOSTRATE: A play there is. . . .
>
> (V. i. 58–61)

The comedies describe play-like worlds so that reality can be viewed apart from life. The imaginary world and the real world are easily separated and imagination serves merely to express passions, not to perceive reality.[30]

The comedies dramatize the ceremonial world of nonpersonal conflict. The ideals of an ordered world which secure the established society-self, the public-private, relationship come under attack, but reconciliation ensues. The dream-stupor of *A Midsummer Night's Dream* and the Arden Forest in *As You Like It* pretentiously cloak man's inevitable conflict with time and its handmaidens, mutability, transience, valuation, and history. Here ritual demise is halted and the maturation of an alternative but isolating individual consciousness is resisted.

A Midsummer Night's Dream focuses on the illusions of ideal love. Any but idyllic love reaps disillusion and frustration. Real love projects the lover quickly into time and into the historical world of his own making.

> Before the time I did Lysander see
> Seemed Athens as a paradise to me
> O' then, what graces in my love do dwell,
> That he hath turned a heaven unto a hell!
> (I. i. 204–207)

Real love, like any nonritually defined action, is most characteristically described in time and consequently its evanescent quality, "Making it momentary as a sound / Swift as a shadow, short as any dream" (I. i. 144–145), belies the ceremonial order and the immutable nature upon which ritual, rather than personal, identity resides.

In the comedic world of the play, then, the dramatic protagonists are not individuals but are forces such as ideal love and passion, or illusion and reality. Indeed, the characters are so lacking in any individual sense of identity that only physical peculiarities mark one from the other. Helena and Hermia are distinguishable because one is taller than the other. Neither conditions the world around her nor motivates the actions which energize this world. Only Bottom and the other artisan actors self-consciously attempt to create something new by naming it or by acting it. Yet their world is but a masque in which reality has been suspended, and while they appreciate the nature of a nonimaginary world—"And yet, to say the truth, reason and love keep little company together nowadays" (III. i. 144–145)—they personate the confounding of illusion and reality. They provide release for the court just as the comedy provides release for Shakespeare's society. "The clowns provide a broad burlesque of the mimetic impulse . . . which in the main action is fulfilled by imagination and understood by humor."[31] As the fairies gather at the end to bless the reconciliating marriage, Puck tells the audience that they have experienced the benefits of slumber and dream, too. Illusion self-consciously denies conflict and reality.[32]

The world of *As You Like It* resembles the dreamworld of *A Midsummer Night's Dream*, but the challenges to the ceremonial and ritual world of order are more vital here and the final reconciliation is more magical, more self-conscious. Here, too, for all intents and purposes, distinguishable identity can be determined by physical features alone. Rosalind owns no final inner self-concept which she acts upon and by which she is known. Indeed, by a change of clothes, she can disguise herself—even from her lover, though she claims that her more than common height is advantageous here, too.[33]

So, too, the forest world here, like the dreamworld in the earlier play, is pretentious; it suggests an ambience in which personal conflict and historical definition do not exist. Only the wind and other natural forces mark discomfort. The forest seems to provide the characters with a nonexistential world in which the rites of love can be actualized on the transcendental level of ideal ritual.

But even in the forest ceremonial ideals must be examined, and a disguised Rosalind non-foolishly plays the role of an objective observer, self-consciously commenting on her own love experience. Again, as in the earlier play, varieties of love are the dramatic protagonists. Corin wants to believe in the total and final definitive constancy in love, but Touchstone counters with the fools' wisdom: "but as all is mortal in nature, so is all nature in love mortal in folly" (II. iv. 52–54). Love can be idyllic or it can be sensual and passionate as Rosalind and Touchstone make clear, but the true arbitrator of the dramatic conflict is time. First we ripe, then we rot, Jaques informs the Duke, "And thereby hangs a tale" (II. vii. 28).

Since in disguise she can objectively, and as it were, self-consciously appraise her own love conflict, Rosalind experiences time's reality. Her first question to Orlando, "I pray you what is't o'clock" (III. ii. 266), reveals her stubborn self-consciousness even within the pretentiously ahistorical forest. Time gives the lie to false ideals which require a constant and immutable nature to sustain them.

> Men are April when they woo, December when they wed. Maids are May when they are maids, but the sky changes when they are wives.
>
> (IV. i. 140–142)

Rosalind appreciates the relativity of experience just as she realizes the variety of its nature.

As You Like It ceremonially concludes as Hymen descends to "bar confusion" and to preside over Rosalind's marriage. Rosalind understands courtly love and experiences passion. Her marriage, however, suggestively reconciles any potential conflict or contradiction therein. Ceremony stabilizes social identity and contrives against the private choices and the burden of self-conscious responsibility that the play both brings to life and buries. After the marriage all conflict is apparently resolved. The old order reaffirms itself. As the Duke tells Celia:

> O my dear niece, welcome thou art to me
> Even daughter, welcome, in no less degree.
>
> (V. iv. 147–148)

Arden can be abandoned now. An artificial concord is erected, and a trip to never-never land postpones a journey into history.

In the *Richard II–Henry IV* cycle of history plays, Shakespeare begins to dramatize the conflict between public and private identity. Here as in the comedies, such conflict takes up residence on a level beyond the truly personal. Now social institutions are examined just as was ceremony in the comedies. Yet "Time" remains the final arbitrator. Confronted by the conspicuous decay of the social order, social identity becomes precarious, and individuals face mounting pressure from private desires which suggest an alternative understanding of self.

Richard II investigates the order-disorder and the public-private antagonisms by focusing on the nature of kingship. It becomes quickly apparent that Richard cannot reconcile them. While the rebellious Bolingbroke challenges the social order and the nature of kingship, it is equally clear that Richard himself threatens both by his behavior, too.[34] Indeed, throughout the play the Dionysian metaphor of fire, which usually expresses disorder in Shakespeare's works, describes them both, and they take on a reciprocal sacrificial relationship, much in the way that Caesar and Brutus do in *Julius Caesar*. Time confounds sacrificer and sacrificed and confuses order and disorder. No distinctions exist.

Richard is the dramatic foil for Richard II as the play investigates the ritual nature of kingship. Richard tries to exercise the ceremonial ritual magic that kingship should provide, but fails, and what this failure makes manifest, on the contrary, is the human mortal nature of the king.

> for within the hollow crown
> That rounds the mortal temples of a king,
> Keeps Death his court, and there the antic sits,
> Scoffing his state and grinning at his pomp,
> . . . . . .
> Mock not flesh and blood
> With solemn reverence throw away respect,
> Tradition, form, and ceremonious duty,
> For you have mistook me all this while:
> I live with bread like you, feel want,
> Taste grief, need friends—subjected thus,
> How can you say to me, I am a king?
> (III. ii. 160–163, 171–177)

Can a man not be a king? There is social as well as personal tragedy in the awareness that the king is more man than God. What magic resides in an institutional identity that is so prone to the demands of mortality?

Upon Richard's struggle to deal with the double aspect of his identity rests both the personal and the social sensitivity of the play. He recognizes his two selves, but his private self informs both Richards. Multidimensionality hounds him: and in a world in which unity defines security and order, Richard's still immature consciousness of self and of "other" self circumscribes the beginnings of a tragic conflict.

> O, that I were as great
> As is my grief, or lesser than my name!
> Or that I could forget what I have been!
> Or not remember what I must be now!
> (III. iii. 136–140)

Richard wishes that he were either a great king or an ordinary man—but he is an ordinary man who is a king. He can resign his title; he knows he cannot lose himself.

> BOLINGBROKE: I thought you had been willing to resign.
> RICHARD: My crown I am, but still my griefs are mine.
> (IV. i. 190–191).

After disavowing his public identity and disowning his kingly acts, Richard confronts the challenge of securing his identity in himself. Yet he objectively and self-consciously views himself as a traitor to his king, to his public identity, and he struggles under the burden of such a loss. He experiences the gap between self and identity.

> I have no name, no title; . . .
> And know not what name to call myself!
> O, that I were a mockery king of snow,
> Standing before the sun of Bolingbroke,
> To melt away in waterdrops!
> (IV. i. 255, 259–262)

If he could be a Lord of Misrule, like Falstaff, he could express his private self; he could attack order and ceremony, but not lose his bearings. Richard wants to avoid the tragic burdens of his own actions. He would like to be a comic figure—and here he foreshadows Falstaff who will melt before Henry's sun.

Alone, contemplating the nature of order (V. v. 1–65), Richard recognizes that his thoughts are creative; they populate the prison as people fill the world. And consequently, all worlds are self-defined. But as he knows, self has more than one meaning, more than one identity, just as thoughts do. And man's greatest thoughts and desires, his ideals, constantly wash upon the shore of real existence. Frustration, then, ensues, and the burden of one's action, in time, proves tragic.

> But whate'er I be,
> Nor I, nor any man that but man is,
> With nothing shall be pleased, till he be eased
> With being nothing. . . .
> (V. v. 38–41)

So no matter how many faces or identities a man assumes, Richard realizes, he is finally and ultimately alone, empty, nothing. Yet such a stark truth demands an acceptance of nothingness as a prerequisite for identity and is consequently, an early essential experiencing of the tragic nature that later Shakespearean heroes such as Cordelia and Coriolanus act out. It is, indeed, an embryonic awareness of the historical and the psychological roots of social meaning and order.

Richard admits that he was out of touch with the social order. And he recognizes also that dis-order results from ignoring time's true nature.[35]

> But for the concord of my state and time
> Had not an ear to hear my true time broke.
> I wasted time, and now doth time waste me.
> (V. v. 47–49)

One must serve time as one's God to keep in touch with change, with history. And in his last moments, Richard, as did Faustus, records time's passage as the illusion of immutability existentially fades for him.

Richard's recognitions, while still only theoretically experienced, represented a challenge to the same order he had lost. And it is, then, not surprising that his end should prepare us for the end of *Henry IV, Part II.*

> Great King, with this coffin I present
> thy buried fear.
> (V. vi. 30–31, PII)

In the Falstaff histories, Shakespeare works "out attitudes toward chivalry, the state and crown in history, in response to the challenge posed by the fate he had dramatized in *Richard II*."[36] Here, too, social institutions embody a dis-eased social order. Again identity is precarious, and time haunts consciousness as if every change it wrought produced death. Richard lives in a world where ritual fails to sustain the magical identity it describes.[37] Falstaff dramatizes the private source of such ritual demise, and constantly seeks to create identity anew by exercising those personal energies. Yet he attains no final, secure identity. His buffoonery and folly exaggerate a still unrefined self-consciousness. Buffoonery becomes a defense for a nascent individuality.

From the very beginning of *Henry IV, Part I*, Falstaff identifies himself with time. Though Hal may question "What a devil hast thou to do with time of day?" (I. ii. 1–2, PI), it soon becomes apparent that Falstaff has everything to do with time. Indeed, as the embodiment of the dis-ease and the dis-order which frustratingly challenge the existent social institutions, Falstaff incorporates change, movement, and time. Though his behavior may appear to some as untimely, he never resists change nor blames time for the state of social disease. As he notes in *Part II*:

> I do see the bottom of Justice Shallow
> Lord, Lord, how subject we old men are to this vice of lying
> . . . Let time shape, and there an end.
> (III. ii. 302–304, 331, PII)

His self-consciousness is buoyed by his recognition that time denies fixed definition. But his acceptance of the flux of things also helps him to resist the burden of his own actions and ultimately inhibits him from pretending any secure identity though he tries on a variety of names and personae throughout. As "time's fool," Falstaff represents life unrestricted by ceremony and ritual, and flows along passively with change. But in *Henry IV* time becomes inextricably identified with judgment, as the traditional institutions of social justice and of redemption continue to decay. And when Falstaff steps out of time and tries to fix as permanent the experiences from *Part I* in *Part II*, time "must have a stop," and judgment reigns.

Although Hal and Falstaff play together in a night world of disorder, neither can define a new internal personal identity capable of sustaining positive action in the public world. But while Falstaff rejects the empty rituals and ceremonies of the old order, Hal slowly reinvokes them by emulating Percy:

> For my part I may speak it to my shame;
> I have a truant been to chivalry.
> (V. i. 93–94, PI)

Because Falstaff appreciates the temporal quality of life, he emancipates himself from false and unrewarding ceremonial cant as Percy and Hal cannot. He respects life and does not make of it a medium for contrived actions as does Percy whose chivalric identity is so precarious that he behaves as if a permanent comatose state will envelop him after a moment's rest. So when Falstaff returns, after feigning death to preserve his life, Hal's surprise reflects Falstaff's direct challenge to his freshly revitalized identity. As Hal remarks: "Thou are not what though seem'st" (V. iv. 136, PI). Indeed, Falstaff is neither dead nor a fool. And for both reasons he threatens Hal who learns that things are not absolutely valued and judged but ignores this lesson and seeks to secure identity with just such "seemings."

Although Hal represents new hope for the institution of kingship and for the social order as a whole, he rejects change and conflict, and immediately upon assuming his duties he identifies himself with his old nemesis, the chief justice. This symbol of an old law and a decrepit order negates whatever potential Hal once possessed for directing social change. Hal negates his past—calls it a dream. But it is with a comedic type of dream-illusion that he confronts Falstaff. Hal's perceptions change. Unlike Hamlet, who struggles to transcend certain "seemings," Hal regressively trades the perceptive imagination of night magic for the worn-out magic of ceremonial ritual. To Falstaff he declares:

> I have long dreamed of such a kind of man,
> So surfeit-swelled, so old, and so profane;
> But, being awaked, I do despise my dream.
> Make less thy body hence, and more thy grace,
> Leave gormandizing, know the grave doth gape
> For thee thrice wider than for other men.
> (V. v. 50–55, PII)

In banishing Falstaff, Hal tries to banish conflict. But the ritual magic herein attempted—the desire to repress imagination, change, and conflict by burying their physical symbol with empty words—is troublesome. By trying to revalidate ritual instead of continuing to dramatize its ineffectiveness, Shakespeare plays with time.[38] Again reconciliation deceptively resists the internalization of social conflict.

Yet tragedy is existential, "something infinite inprisoned in the finite,"[39] and in the early problematic tragedies, *Julius Caesar, Troilus and Cressida*, and *Hamlet*, Shakespeare begins to dramatize the process of the internalization of social conflict. The sacrificial energy in tragedy directs this process and fashions an effort that seeks to transcend the limiting character of existential reality, which the demise of social order and identity and a concomitant development of self-conscious redefinition reveals. It is only by a working through and an eventual failing of this tragic vision that one makes "meaningful" this process of internalization. As Frye says, we "come to terms with irony by reducing our wants."[40]

The questioning of ceremonial and institutional order in the early comedies and histories parallels the rise of extreme doubts concerning social identity in the early tragedies. "The feeling of lost social identity is what is expressed in the story of the fallen prince."[41] In the face of this loss and the consequent uncertainty about the meaning of existence, Brutus, Hector, Achilles, and Hamlet each struggles to find an alternative fixed desideratum with which to judge actions and from which then to derive security and identity. "The ordered society in Shakespeare is, to use Heidegger's terms, ecstatic: its members are outside themselves, at work in the world, and their being is their function."[42] And it is surely as an examination of the problematic nature of this failing order that Shakespeare's vision dramatizes the process of the development of self-consciousness.

Meaning in the early tragedies is "susceptible of contradictory interpretations" while "consciousness . . . is almost inevitably complementary."[43] Whether one's actions reflect private or public values and whether value itself is ascribed or achieved or perhaps is inherent in an action or a thing—such concerns repetitively shape the dramatic tension in these plays. Throughout all of this the dramatic relationship between meaning and self-responsibility clarifies the introspective nature of the tragedies. And it is only through such a continued struggle with the problems of meaning and identity in these early tragedies that in the later tragedies Shakespeare's vision begins to focus upon the individual, the isolated self, as the ultimate source and medium for meaning and order in the social world.

"*Julius Caesar* is about the problems of spiritual economy—of dealing with one's self."[44] Brutus suggests an extremely introspective Richard II. He burdens himself with his passions and problems and early in the play hints that he possesses the capacity for self-knowledge and perhaps, too, the potential for self-defined identity. Yet as did Richard II, Brutus requires a mirror, some "other," to reflect back to him his self.

> CASSIUS: Tell me, good Brutus, can you see your Face?
> BRUTUS: No, Cassius; For the eye sees not itself
> But by reflection, by some other things.
> (I. ii. 51–53)

Cassius offers his services as Brutus' "other," his mirror, to mediate between Brutus and Brutus' "self":

> I your glass
> will modestly discover to yourself
> That of yourself which you yet know not of.
> (I. ii. 68–70)

But it remains evident that Brutus resists looking into his private self for identity. Early in the play, then, the inevitable link between self and other is established. What needs yet to be determined, however, is their definitive relationship.

Throughout, Brutus nurtures his public identity as he maintains a stoical private self. Like his alter ego, Caesar, he surrounds himself with ceremony and magic and like Caesar, too, he invokes the power of his public identity to justify his behavior. From the beginning Brutus seeks security and meaning in his public "honor" rather than from private motivations.

> I love
> The name of honour more than I fear death.
> (I. ii. 89–90)

Caesar, too, relies upon his public self and invokes the power in his name much as did Richard II (I. ii. 211–212). And even Cassius, who most of all recognizes man's responsibility for his own fate, plays with names in a way which establishes the repetitive, ceremonial motif that continues throughout the play:

> Brutus and Caesar: what should be in that name 'Caesar'?
> Why should that name be sounded more than yours?
> (I. ii. 142–143)

Confronted by his own imagination and by his own private actions, Brutus retreats from an acceptance of self-consciousness and attempts to cloak himself and his acts in a ritualized, ceremonial guise which smacks of medieval magic. He refuses to oath or contract with the other conspirators but tries rather to pretend a magical bond consisting of words and names which would fix immutable their wills as action and assure them of everlasting meaning.

> what other bond
> Than secret Romans that have spoke the word,
> And will not palter? and what other oath
> Than honesty to honesty engaged.
> (II. i. 124–127)

And later when he explains the murder to the Roman people he does not attempt to justify his deeds or actions, but rather he again invokes the power of his public identity.[45]

> believe me for mine
> honour, and have respect to mine
> honour, that you may believe.
> (III, ii. 14–16)

But Brutus is involved in a magical world that extends far beyond the boundaries of ritual invocation. To the sick, dis-eased social body he offers his services as exorcist, and beyond this, too, he promises a magical world of self-definition, indicating meaning and order based upon a positive relationship between man and nature. As Ligarius tells Brutus:

> I here discard my sickness!
> Soul of Rome!
> Brave son, derived From honourable loins!
> Thou, like an exorcist, hast conjured up
> My mortified spirit. Now bid me run,
> And I will strive with things impossible.
> (II. i. 321–325)

Sickness and disorder are transcended in the sacrificial magic of the assassination too. The conspirators surround Caesar like a pack of wolves circling their prey. The sacrificial murder reverses the wolf-sheep metaphor and suggests political emancipation.[46] On another level, though, the conspirators seem to form a conjuring circle around Caesar as if by sacrificing him they could exorcise social passions and social distemper and summon back from the dead past social order. "The desired effect of the sacrifice (wholeness, continuity) is replaced in the sacrificial act itself by its cause (a felt lack or void); the sacrificial victim is voided, discontinued, so that there will be fullness, wholeness, continuity."[47] And so too at Caesar's funeral: "Then make a ring about the corpse of Caesar" (IV. i. 159). Caesar the sacrificed and Brutus the sacrificer together invoke the ritual magic of the past, exorcise the passions of social disorder, and finally, by sacrificing themselves in the name of social order, unself-consciously lead society beyond the ceremonial ritual of the old order and into the magic of self-definition.[48]

Julius Caesar cautions against the temptation of absolute, of final or fixed valuation. Against such a vision consciousness must wage a tragic battle before it can utilize self-conscious responsibility and self-definition in the cause of social order. Naming something does not free that thing from other meanings or from time's alterations. Antony refers to Caesar as the "noblest man" and later to Brutus as the "noblest Roman." How meaningless such naming proves to be, how inflated language has become. Antony further debases the value of such accolades by claiming that Cassius is "A prize no less in worth" than is Brutus.

Brutus' convictions prove to be just as empty of finality, just as illusory. His description of the "sacrificial" nature of Caesar's death is realistically refuted by Antony's "butchers" description. And finally Brutus considers the nature of reality:

> Why then, lead on. O, that a man might know
> The end of this day's business ere it come!
> But it sufficeth that the day will end
> And then the end is known. Come, ho! away!
> (V. i. 122–125)

As usual, his contradictory consciousness accepts that he can define his own victory and fortune, but only within the limitations of some other defined cyclical nature of things.

> We, at the height, are ready to decline
> There is a tide in the affairs of men
> Which taken at the flood leads on to fortune;
> Omitted, all the voyage of their life
> Is bound in shallows and in miseries.
> On such a full sea are we now afloat,
> And we must take the current when it serves,
> Or lose our ventures.
> (IV. iii. 215–222)

Troilus and Cressida, more than any other of Shakespeare's plays, emphasizes the concern with meaning and valuation, or perhaps with devaluation. "As soon as one turns too sharply back on what one is, the self dissolves, or, conversely, if one too ardently equates one's self with any available codified description, the act itself falsifies and kills."[49] The dramatic focus of the play centers on the inevitable relationship between self and other which a search for certainty and order makes requisite. Man's identity takes form and becomes known in communication. It can not be formed, known, or sustained in separation from society or others.[50] Consequently, *Troilus and Cressida* investigates the complex nature of truth, identity, and self, and subsequently of order, too. To see this play, as many have, as a celebration of order is to "confuse the active problem with its traditional answer."[51]

Although throughout the play character after character seeks answers to questions about truth and value, from the beginning "Tarry" is the word. As Brutus finally realized, time will tell and value must only be appreciated as a function of time.[52] Inconstancy and mutability naturally give the lie to all attempts to fix value in self through other. Yet everything gains meaning only in relation to other things. One's sense of self depends upon other, while one's understanding of other reflects one's self-appraisal.

> to end a tale of length
> Troy in our weakness stands, not in her strength.
> (I. iii. 136–137)

Troilus' relationship with Cressida dramatizes the existentially impossible nature of intrinsic value. Self-integrity melts away in the course of social interaction. Activity is fluidly expressive as is time itself, but the "Act" is final and alters the meaning of the action as Hamlet knows. Becoming, not being, denies the definitive one act. As Cressida observed: "Thing's won are done—joy's soul lies in the doing" (I. ii. 288).

Cressida self-consciously defends her self by maintaining as fluid a self-definition as she can. More mature than Falstaff, her chameleonic personality protects her against those who try to limit and define her actions. She is well aware that Troilus' valuing of her is not in accord with her worth but rather reflects his needs. She attempts, then, to actualize her self within the context of social expectation. And she appreciates the mutability therein:

> I have a kind of self resides with you,
> But an unkind self that itself will leave,
> To be another's fool.
> (III. ii. 147–149)

Troilus, on the other hand, identifies his self with truth and truth with simplicity. He neither appreciates the complex nature of self-definition nor the complementary nature of value and truth ("true and not true"). He metonymically reduces himself to "True Troilus" (Honorable Brutus) and consequently sacrifices his capacity for self-definition. Cressida recognizes Troilus' misconception and dramatically teaches him about the dangers of such simple and limiting absolutes. "As false as Cressid'" becomes an incontrovertible truth.

Later, in the Greek camp Cressida and Diomedes meet. From a distance Troilus, along with Ulysses, watch. And from a more profound vantage yet, Thersites watches the lovers and watches the watchers. Awareness takes on a spatial quality here which keeps truth at a distance and maintains its relative nature. Troilus refuses to face up to what he sees. It is his tenuous self-identity which suffers for Cressida's infidelity. And he tries to deny the complementary complexity of the truth.

> This she? No; this is Diomed's Cressida.
> . . . . . .
> If there be rule in unity itself,
> This is not she . . .
> . . . . . .

> Bifold authority! where reason can revolt
> Without perdition, and loss assume all reason
> Without revolt. This is, and is not, Cressid!
> Within my soul there doth conduce a fight
> Of this strange nature, that a thing inseparate
> Divides more wider than the sky and earth.
> (V. ii. 137–149)

Indeed Cressida is someone else to everyone, just as she knew. Identity can only be negotiated in social intercourse. And the consciousness of such an awareness is requisite for the positive exercising of self-definition. Cressida's complex self forces Troilus to confront the existential nature of reality. There is no rule in unity. Simple definitions, whether externally or internally derived, can not deny time's mutable signs. There can be no simple alternative to the demise of an ordered cosmic vision. Meaning redefinition is a cultural process. Identity and ideology become through discursive negotiation. Time governs history, and truth and value indicate this temporal process. This is and is not Cressida as it is and is not true. The emancipated self struggles to resist absolute imperatives of all derivations.

Troilus' confrontation is personal as well as political, private as well as social. It is of "Fall-like" proportions. But the "Fall" now is into the self-consciousness of consciousness. And here the poetry captures the feeling of the original primal battle between Mother Earth and Father Sky and the subsequent castration. Division ensues from one into many; from unity into variety; from eternity into history; from peace into politics and sublimation; and from cosmic order into personality and self-definition.

Yet what is self-definition? How does it express itself? Does any Shakespearean character suffer with this dilemma more than Hamlet does? For him, everything is merely "seems," but "seems" is not self. As he tells his mother early in the first act about various expressions and actions:

> These indeed seem,
> For they are actions that a man might play,
> But I have that within which passes show,
> These but the trappings and suits of woe.
> (I. ii. 83–86)

He denies that his true self can be acted out or expressed. Appearances are just obfuscations. Hamlet suggests that there exists within self a deeper, more profoundly insular truth which neither actions nor speech can represent. But he does not easily find a positive way of representing himself in and to the world. As Erich Heller notes:

> Hamlet is the man who has bequeathed to modern literature and thought the obsessive preoccupation with "authenticity," the much-desired virtue that calls every other virtue hypocrite or, more fashionably *mauvaise*, bad faith. Of course authenticity, in this sense is a chimera . . . because there is for Hamlet nothing—nothing he might do or not do—that could possibly be in perfect accord with his inner being.[53]

Hamlet goes beyond *Julius Caesar* and *Troilus and Cressida* in that it rejects the real possibility of public identity and denies, too, a viable sense of self which derives from social intercourse and is dependent upon a relationship with other. While *Hamlet* dramatizes the profound struggle to identify the nature of self-definition it tragically resists any attempt to establish a positive self-other relationship.[54]

For Hamlet every relationship illustrates a tension between some private sense of self and an interfering public demand, between, as Eagleton puts it, the self as subject and the self as object.[55] Where communication established meaning in *Troilus and Cressida*, in *Hamlet* it leads to doom. By giving his "ear" to the ghost, Hamlet becomes the object of an external subjective experience. The outcome of such a relationship is one of subjugation. Hamlet feels driven to revenge by an external source. By committing the revenge murder Hamlet would lose his self—as indeed he literally does. So too does Laertes give his "ear" to Claudius with similar results. And more literally still does Hamlet's father give his "ear" to the poison of Claudius. Polonius' advice—"give every man thy ear"—is blatantly dangerous, then. *Hamlet* maintains that a true and positive source for action and self-expression cannot be sought or found outside of self. Hamlet's two possible actions—princehood and murder—both reflect an externally defined purpose and he resists them both for as long as he can.[56] But in doing so he does not act at all and he sustains his self in isolation, by denying himself an expressive outlet. It is in light of this that his disowning of appearances and actions must be appreciated. The complexity of the true self confounds the limiting simplicity of the particular defining action. Can it express itself, however, outside of itself?[57]

Hamlet realizes the true nature of his tragedy. Denmark is his prison because while he "seems" to be free to act upon his will he knows that all the paths open to him have been predetermined. His tragedy expands his consciousness of self and forces him to confront existential reality. There can be no escape.

> O God! I could be bounded
> in a nut-shell, and count myself
> a king of infinite space; were it
> not that I have bad dreams.
> (II. ii. 257–259)

Yet his awareness of this irony leads to doubt about the nature of self-expression and eventually to an appreciation of the human potential. It is only when one's received order is challenged by "bad dreams" or imagination that self-integrity and the authenticity of self-definition become the focus of consciousness. In Shakespeare's vision it is the self-consciously author-itative step in the redefinition of meaning and order.

The famous soliloquy (III. i) captures Hamlet's introspective struggle with the true nature of self. Hamlet ponders more than the relative merits of life and death; he questions the quality of life as a creative self-defining force. "To be"—is to be authentic as self. Hamlet measures this against "not to be"—to become. Can one *Be* self or only *Become* self? Can self only be known and expressed potentially? Indeed can there *Be* self in time, in history? Can self exist within a social context? Is it possible to finalize self-potential in a creative, expressive Act? Or is the only finally expressive Act the ultimate negation—suicide? Must one end one's life in order to cement the gap between authentic self-expression and the concrete sterility of the action's afterlife? Is this the context in which Hamlet charges his mother to "Repent what's past, avoid what is to come" (III. iv. 150)? Will one act defy time and contain the past, the present and the future within itself? Does Hamlet expect just a word of repentance here, or, rather, a final self-negating action?

As Hamlet continues to resist his inevitable course of action he becomes steadily surer that any who pretend to know a man by his actions know little more than his fleeting shadow (V. ii. 123–124). Even when he spurs himself after comparison to Fortinbras and self-consciously directs himself to plan and to manipulate future events, the best he can conclude is:

> O, from this time forth,
> My *thoughts* be bloody, or be nothing
> worth.
> (IV. iv. 65–66, emphasis mine)

Ironically, perhaps, he prepares to act spontaneously and authentically when the time will demand it and with Caesar, Hector, and others he defies augury and declares that "the readiness is all" (V. ii. 120). Preparation, then, is self-expressive, and conceptual reality reflects the self better than action does. Indeed for Hamlet, words are like acts too, and do not reflect self because they contain meaning-in-themselves. Self-conscious definition confronts a world without absolute standards and in which meaning continually changes over time. Hamlet boldly and radically attempts to conceptualize authentic meaning and identity because he realizes that

> there is nothing either good
> or bad, but thinking makes it so.
> (II. ii. 252–253)

In *King Lear*, in *Macbeth*, and in *Coriolanus* Shakespeare's tragic vision dramatizes alternative responses to Hamlet's existential dilemma. Cordelia, the Macbeths, and Coriolanus look within for a way out of a world which offers no final order for existence and no fixed purpose or meaning against which to judge behavior. Each seeks security and certainty in an isolated, self-defined, authentic individuality. For each the self becomes a final negative medium for truth and meaning and leads ultimately to social irresponsibility and tragedy.

The crux of *King Lear*, and perhaps the central moment of Shakespeare's entire tragic vision, is Cordelia's response to her father's mock ceremonial question concerning her interest in winning an "opulent" portion of his public realm. Lear has already stripped himself of his public responsibilities and burdens before he curiously asks his daughters about their personal love for him. Selfishly, he renounces his social obligations by parceling up his public lands, and ironically asks his daughters to give of their "personal" selves and to accept his public duties as well.

Cordelia's haughty, yet poignant response to her father's inquiry about what she might say to win more land,

> Nothing, my lord.
> . . . . . .
> Unhappy that I am, I cannot heave
> My heart into my mouth, I love your Majesty
> According to my bond, no more nor less.
> (I. i. 86, 90–92)

concentrates Shakespeare's entire tragic vision and firmly links Hamlet to Coriolanus. As Hamlet knew, words like actions are external to the self and do not express or reflect self. Love as being can be most fully expressed in readiness, and as was Hamlet, Cordelia is afraid of subjecting readiness and consequently her authentic self-expression to words or actions. By resisting her father's request, then, she maintains her freedom, her self-definition; but as is true for Coriolanus and Hamlet, too, the price of such integrity is an ultimately negative self-expressiveness.

Upon her father's further inquisition Cordelia maintains that she is "true." Lear's response draws attention to the isolated nature of such self-truth. "Let it be so; thy truth then be thy dower!" (I. i. 107). Let truth be your value, he says. Can you deal it or deal with it? Is it useful? A dowry is one of the means by which one authenticated self socially and defined a reciprocal self-other relationship. As such a medium of exchange, Lear suggests, self-truth will not prove valuable.

Yet Lear can learn from Cordelia's behavior, and the fool undertakes to help him: "Can you make no use of nothing nuncle?" (I. iv. 132). The fool tries to teach Lear to resist strict absolute definitions which limit and preordain experience. He rails at him not to fix things certain but to be fluid like nature (I. iv. 119–128). And he continually expresses an oxymoronic, complementary vision which denies any secure, literally reduced sense of things. Lear experiences confusion on all sides and is left, himself, with nothing. (I. iv. 186–188). Stripped of all the accoutrements of his ceremonial identity, Lear follows the fool into nature and simultaneously looks into himself for his own nature, too.

The storm and his wanderings in the elements transform Lear so that he, indeed, begins metaphorically to understand the flux that is nature. When he at last sees the oxymoronic sense of things, the fool's job is done and he disappears (III. vi. 82–84). Nature and flux teach Lear that the desire for proof, meaning, cause, or purpose beyond action itself is superfluous and finally unnatural.

But there can be no reconciliation. The sacrificial energy of the tragic vision must work itself out. As did Hamlet and Coriolanus, Cordelia sacrifices self-nature. While the Caesar–Brutus and the Hector–Achilles sacrifices exorcised the social malaise that accompanied the decline of the old order, Cordelia's sacrifice, like those of Hamlet and Coriolanus, begins to exorcise from the self-conscious individual the negating and isolating nature of authentic self-identity. *King Lear* concludes desperately. Order is not facilely restored and no promising youth suggests a better future. But self-sacrifice indicates a socially responsible means to deal with the negative experience of an isolating self-consciousness.

Just as in *King Lear*, the confounding of predefined certainty and absolute evaluation by handy-dandy and oxymoron marks *Macbeth* from the outset (I. i. iv, 11). Into this world steps Macbeth,

> Disdaining fortune, with his brandished steel,
> Which smoked with bloody execution,
> Like Valour's minion carved out his passage.
> (I. ii. 17–19)

He has made his own way in the world and Shakespeare centers in him another dramatization of the possibilities and consequences of total self-definition.

Macbeth desires "action without consequence"—fully self-contained action.[58] He hopes to establish absolute meaning for life in one act of self-definition. He aspires after kingship but he resists opening himself to the "fluid, temporal process of actions necessary to win and secure it."[59] Lady Macbeth encourages him and envisions that they can step out of time and can act to transcend both the fluidity and process of existential experience and received meaning.

> . . . and I feel now
> The future in the instant.
> (I. v. 56–57)

Macbeth's aspirations and actions direct attention toward the relationship between self and other. He seems to appreciate the necessary social nature of self.

> I dare do all that may become a man;
> Who dares do more, is none.
> (I. vii. 46–47)

To be more than a man would be to negate self because man's nature is social as well as private. But Lady Macbeth urges Macbeth to contemplate transcendence.

> And, to be more than what you were, you would
> Be so much more than man.
> (I. vii. 50–51)

She confuses the isolated self with the aggrandized self and in so doing negates the creative possibilities in tragic consciousness. She pits self-consciousness against life rather than attempting to celebrate its temporal nature. Finally they agree to flaunt time and to act authentically and absolutely. As Macbeth says,

> I am settled . . .
> . . . . . .
> Away, and mock the time with fairest show:
> False face must hide what the false heart
> doth know.
> (I. vii. 79–82)

But his willingness to bear false face and false heart reveals the negative, isolated quality of his action. To be true to self Macbeth must be false to others and must then negate his identity. And if the face and the heart are false what, indeed, can finally reflect authentic self except negation?

Macbeth's tragedy resides not in the nature of his actions alone, but also in the impossible ends to which he directs them. "Let every man be master of his time," he says (III. i. 40). He hopes to eternalize a moment with a single act and he sacrifices his entire self to that purpose. Yet even though he succeeds in winning a kingship he learns that Banquo's heirs will succeed him in time. His act, irrespective of how self-consciously he energized it, accomplishes no more than itself, and time, not human action, measures existential reality.

The witches can do what Macbeth can not: "A deed without a name" (IV. i. 48). They create; but they resist literalizing and limiting the meaning of their actions. By isolating himself from positive social intercourse Macbeth absolutely limits the nature of his potential self-definition and fails to attain the security and identity he desires.

With *Coriolanus* Shakespeare's tragic vision reaches an extreme but logical conclusion. More than any other hero in the canon, Coriolanus attempts to locate meaning and definition in a constant, private self. No other play dramatizes such a complete antagonism between "authentic self-expression and social responsibility."[60] And Coriolanus' confident self-consciousness, the medium through which he acts to define meaning and identity, develops and sustains itself at the expense of any external motivation or public purpose.

Coriolanus conspicuously separates self from other, private from public. He trusts neither other individuals nor the public realm which supposedly legitimates identity. Against the flux of existence he maintains: "I am constant" (I. i. 238). He shuns association with the people who reflect the inconsistency that he so terribly dreads.

> . . . He that trusts to you
> Where he should find you lions, finds you hares;
> Where foxes geese: you are no surer, no,
> Than is the cool of fire upon the ice,
> Or hailstone in the sun . . .
> . . . . . .
> Trust ye?
> With every minute you do change a mind . . .
> (I. i. 169–173, 180–181)

And he resists linking his actions to a public, civic cause whose embodiment, Rome, no longer expresses consistency or truth. Coriolanus vents his disillusion and disgust with the inauthenticity of public purpose on those who seek to secure identity socially.

> I would they were barbarians, as they are,
> Though in Rome littered; not Romans, as
> they are not
> Though calved i' th' porch o' th' Capital.
> (III. i. 237–239)

Unlike Macbeth, Coriolanus can perform "a deed with no name" and he strenuously resists the attempts of others to "name" his actions and thereby to tie them to some purpose or meaning outside of themselves. As Cominius correctly states, Coriolanus

rewards
His deeds with doing them, and is content
To spend the time to end it.
(II. ii. 125–127)

He knows what Lear and Macbeth struggled to learn: that superfluous explanation and meaning are irrelevant. One's actions exist and in doing them one expresses one's self authentically. But by marrying his security and identity so vehemently to existential action Coriolanus traps his self. Whereas Macbeth sought constancy in an act, Coriolanus needs constantly to act to sustain his constancy. And by demanding that his actions authentically represent his self, Coriolanus isolates himself from the only world in which he can act.

As a soldier and a public hero Coriolanus should have no difficulty in actively fulfilling himself. But he continuously denies the public burdens which he believes circumscribe his behavior. He flaunts the rituals, ceremonies, and authority upon which Roman order and tradition rest. He refuses to enter into any relationship with society which he does not prescribe himself.

Know, good mother,
I had rather be their servant in my way
Than sway with them in theirs.
(II. i. 199–201)

He might stand for "consul," but he will not be subjected to any external purpose or definition. Cominius and his mother implore him to go to the people, to don the wolfskin, and to ask to be their "consul," but Coriolanus worries that in so doing he will lose his private self.

I do beseech you
Let me o'erleap that custom, for I can not
Put on the gown, stand naked, and entreat them,
For my wounds' sake, to give their suffrage: please you
That I may pass this doing.
(II. ii. 133–137)

It is not merely the ceremony that Coriolanus fears. He senses that if he puts on the political toggery he will simultaneously strip his self naked. He cannot suffer the mandated presentation of self. He does not want to enter into a social contract with the people and become, in a very Hobbesian sense, their artificial person, their articulated objectified substitute, their represented will. He understands that the price paid for social order, even a self-consciously defined social order, is the voluntary repression of self-will and he is yet unready for such self-sacrifice.

But Rome makes claim on each of her citizens and although Coriolanus refuses public praise and claims that his actions need no judgment, Cominius tells him:

You shall not be
The grave of your deserving; Rome must know
The value of her own: 'twere a concealment
Worse than a theft, no less than a traducement,
To hide your doings.
(I. ix. 19–23)

Each citizen's actions establish the reciprocal nature and facilitate a meaningful and efficient valuation of the relationship between first and final cause, upon which negotiated social order depends. Isolated action, then, denies society a necessary source of rejuvenation and inevitably negates social order as well as self-definition. Aufidius appreciates that only through social interaction can the self actively sustain itself, and he uses this understanding to overcome Coriolanus' superior strength. As he notes early in the play:

> I would I were a Roman; for I cannot
> Being a Volsce, be that I am.
> (I. x. 4–5)

Ironically, later, Coriolanus will need to be a Volsce in order to find an outlet for his self. As Achilles learned in *Troilus and Cressida*, whereas authentic self-expression negates public meaning and identity one cannot truly express one's self except in a social context.

And it is just such a refined appreciation and consciousness of self that Volumnia attempts to impart to her son. Coriolanus asks her:

> would you have me
> False to my nature? Rather say I play
> The man I am.
> (III. ii. 14–16)

And she answers:

> You might have been enough the man you are,
> With striving less to be so.
> (III. ii. 19–20)

For Shakespeare, as well as for Marlowe, the unrefined energy of the striving individual energized emancipated self-consciousness and prepared man psychologically to redefine social meaning and order. But now, as Volumnia knows, the ability to strive less, to be active and passive, is necessary if one is to help constitute a meaningful social context within which one can continue self-consciously to express oneself. And as Shakespeare's vision, as well as Hobbes' political psychology, makes clear, the authentic self is ultimately artificial. Coriolanus "plays" at being himself. His person is a "persona."

Volumnia chides her son for being "too absolute." She prompts him to compromise his nature, to bend his way and consequently to join in the social covenant. He must dissemble and thereby sacrifice the authenticity upon which his sense of self rests. And he must make this sacrifice because others require it.

> I would dissemble with my nature, where
> My fortunes and my friends at stake required
> I should do so in honour.
> (III. ii. 62–64)

Volumnia prompts him. Yet for Coriolanus, self-conscious dissembling is but the prelude to tragedy. Constancy requires total self-isolation to sustain it, and that path as well as self-sacrifice inevitably leads to self-negation and death.

Coriolanus suffers from a similar type of self-defined tragedy as do Marlowe's heroes. Tying oneself to existential absolutes is as damning and as dislocating as living in a world void of functioning essential redemptive meaning. And often, existential isolation profoundly compounds the anomic experience suffered by those who live during such moments. Without social meaning to sustain "integrity" and self-conscious identity, existential activity can only remain isolating, negative, and tragic.

Coriolanus' tragic end dramatically exorcises the negating, isolating nature of self-conscious authenticity. Like Cordelia, Coriolanus attempts to sustain a totally self-defined existence. As Cominius relates to the Romans:

> 'Coriolanus'
> He would not answer to; Forbad all names;
> He was a king of *nothing*, titleless.
> (V. i. 11–13, emphasis mine)

And yet Coriolanus now acts first as an object of the Volsces and then as a sacrificial object of his country and family.

> O mother, mother
> What have you done? . . .
> . . . . . .
> You have won a happy victory to Rome
> But, for your son—believe it O, believe it—
> Most dangerously you have with him prevailed,
> If not most mortal to him.
> . . . . . .
> I'll frame convenient peace.
> (V. iii. 182–191)

And now Volumnia sacrifices Coriolanus and forces him to abandon his last vestige for self-definition and to assume the burden and the suffering of others. Shakespeare's tragic vision has worked itself through. Where Achilles sacrificed his public identity and chose between two private motives, now Coriolanus retreats from his authentic, private self and chooses between Rome and the Volsces.

> for I cannot be
> Mine own, nor any thing to any, if
> I be not thine.
> (IV. iv. 43–45)

Florizal's love for Perdita in *The Winter's Tale* embodies the perfect image of the reciprocal nature of the relationship between man and society, which Shakespeare's last plays dramatize. Self-knowledge lovingly recognizes other and incorporates social responsibility.[61] Thematically as well as chronologically, the last plays follow the tragedies and "point toward a world where the antitheses of spectator and spectacle, the subject confronting the object, no longer exists."[62]

As do Perdita and Florizal, Miranda and Ferdinand in *The Tempest* willingly share their selves as they create a loving, reciprocal relationship, even within Prospero's overriding scheme of things. As Ferdinand tells Miranda:

> . . . I
> Beyond all limit of what else i' th' world,
> Do love, prize, honor you.
> (III. i. 71–73)

They actively express themselves, but they remain passively in touch with nature and with Prospero's design, too. Actives and passives, subject and object, intrinsic value and extrinsic meaning blend together artfully to create a magical world in which man and nature become one. Francisco relates that he saw Ferdinand in the sea after the tempest.

> Sir, he may live.
> I saw him beat the surges under him,
> And ride upon their backs; he trod the water,
> Whose enmity he flung aside, and breasted
> The surge most swoln that met him.
> (II. i. 112–116)

At one with nature, and exercising his own potential too, Ferdinand incorporates the synthesis of all opposites. And his relationship with Miranda, the incarnation of all beauty and unity,

> But you, O you,
> So perfect, and so peerless, are created
> Of every creature's best
> (III. i. 46–48)

indicates redemption. As Prospero notes:

> Fair encounter
> Of two most rare affections: heavens rain grace
> On that which breeds between 'em!
> (III. i. 74–76)

And it is an all-encompassing redemption. Self and society will be saved together. The last plays create an ambience whose fabric is woven from a "deep connectedness of lives."[63] Private life derives from social order and survives in terms of community, and reciprocally, community is ultimately created and sustained by private lives.[64] This relationship between private and the public profoundly reduces the tragic implications of self-consciousness without negating self-expression and without generating a facile sense of transcendental order either. Indeed, such a magical world resembles Hobbes' sovereign state. The relation between the individual and the social appears epistemologically to have transcended the causal nature of things. It is "the relation of aspects of a single life."[65]

Prospero's art, his beneficent natural magic, gives this vision direction and meaning. More subtle than John Dee, more self-assured than Dr. Faustus, and more imaginative than Francis Bacon, Prospero constructs a social order where artificiality is natural. His final vision which his "imagication" translates into social order makes living a natural art.[66] And the purpose of Prospero's art is the "ordination of civility, the control of appetite, the transformation of nature by breeding and learning; it is, even, in a sense, the means of Grace."[67]

But Prospero has refined his vision on his island. Prior to traveling imaginatively into such isolation Prospero could not conciliate his private studies with his public political obligations: contemplation with action. On the island he self-consciously controls his will and his passions, and purposefully reclaims his political power through the use of art and penitence. Prospero advances beyond Lear and learns to be one with nature and consequently to actualize self-socially. And he manipulates nature's elements tempestuously to educate the others in a manner that recalls Lear's wandering lessons.

After self-consciously making his peace with the public world, accepting social responsibility, and acting benevolently by pardoning the usurpers, Prospero reconciles spectacle with spectators. He addresses the audience directly. It is a first step toward relinquishing the imaginative and artistic vision which he has courted alone.

> Let me not,
> Since I have my dukedom got,
> And pardoned the deceiver, dwell
> In this bare island. . . .
> (Epilogue, 5–8)

As man attempts self-consciously to provide his isolated self with an ordered existence, he artfully or artificially redefines the meaning of reality. This is the vision, in a fashion, that Shakespeare's entire canon dramatizes—the breakdown of the received idea of order, a concomitant developing consciousness of self, the isolation of self, and the attempt to define absolute meaning around the isolated self. Prospero's island imagination artfully illustrates the nature of this potentially creative self-isolation. Isolated, Prospero learned that self required other and that self-expression could not shun social obligation and sustain itself. In this way, then, *The Tempest* may be interpreted as a masque which "pageants" the entire tragic canon and projects Shakespeare's vision—out of time. Played out this way Shakespeare transcends his tragic consciousness without retreating into the comedic world of transcendental reality and received idea.

Prospero's active creativity fuses with the more passive "sustaining design" of nature.[68] A reciprocity between creation and restraint characterizes this world. This is more than just a Hermetic vision of the blending of actives and passives; it is more, too, than the neoplatonic ideal of reconciling action and contemplation. It describes, rather, Shakespeare's positive vision which locates meaning and order within the complementary context of social creativity and individual expression. It parallels, in a nonmaterialistic fashion, the fundamental insights in Hobbes' political psychology. And it fashions the framework for a negotiated social order based upon articulated self-definition and social obligation.

CHAPTER 3

Anomalies and Alternatives in Elizabethan England

> In a word, variety: I can not in one
> word better express the sum of all:
> for indeed I have seen muche, & much
> variety in that I have seene.
>
> Nicholas Breton,
> *Choice, Chance, and Change*

The sentiment stated here represents a theme commonly articulated in all genres of Elizabethan expression. Though Breton, a minor writer, published his dialogue in 1606, he captured succinctly the consciousness of variety and change that developed and matured throughout the second half of the sixteenth century. Certainly an awareness of flux and instability in the world is, in varying intensities, typical in historical consciousness. Yet historical moments arise which can be characterized by the profoundly confrontational stance taken during them in relation to this consciousness of change and variety.[1] Articulate Englishmen, during the Elizabethan years which witnessed England's growing ascendancy in the European as well as in the global picture, engaged those ubiquitously human truths, change and variety, with what seemed to be an ever-increasing necessity and vehemence.

The intense preoccupation with these themes by writers of all kinds during the Elizabethan years is wholly understandable. Identity in Tudor England was nourished by a metaphorical idea of order that posited stasis, unity, and hierarchy as objective reality.[2] Yet the reality of change was everywhere. Much of the intellectual energy of this period was directed toward a reconsideration of the meaning of social reality. Simultaneously, then, Elizabethans suffered from cultural anomie and participated in subjectively reconstructing the meaning of social reality. It was during these frenetic, exciting, and anxious years that the basic commonplaces of the Tudor idea of order were most pointedly articulated. Traditional values concerning authority, mixed government, natural law, and order and degree were reemphasized laboriously as theorists retreated under the emotional impact of a changing world that demanded changing values. The support of the old and traditional was sought as the new and radical grew. Alongside the growing absolutism of the monarch and the developing independence of Parliament, medieval traditions concerning the law and government were continually expressed.[3]

Such articulate defense of the traditional, motivated as it was by the psychological challenge that change assured, necessarily incorporated a developing consciousness, if not an acceptance, of the innovative. Consequently, the Elizabethan confrontation with change and variety, even if outlined as a staunchly conservative plea for the status quo, or even for the ancient customs of a beloved past, in a general and often unconscious fashion, processed the secular and radical values that mediated the examination of the idea of order.[4] No single theorist of any persuasion posited as radical a system of values as did Christopher Marlowe, and perhaps only Richard Hooker, among theorists, articulated the antithetical, yet complementary sets of values and issues that energized Elizabethan consciousness, as did Shakespeare. Nevertheless, Elizabethan theorists, in general, began to outline social issues in a dramatically radical, secular fashion.

Concern with change and variety inspired discussion of time, decay, and historical cycles. The introduction of Italian humanism into England early in the sixteenth century had already established these subjects as intellectual issues. But the growing tensions, social and intellectual, that plagued the second half of the century transformed these originally sterile, borrowed, and Latin issues into emotionally ripe, personal ones. Social, political, historical, and eventually private phenomena began to become temporally and secularly situated. Explanations and judgments that concerned social and political issues emphasized this-worldly considerations. Recourse to religious and scriptural evidence lessened as theorists and practical politicians emphasized the temporal nature of social ills.

A self-conscious awareness and acceptance of change and variety did not immediately and indiscriminately destroy the traditional, otherworldly emphasis of the Tudor idea of order, of course. Early Tudor humanists developed the Renaissance concepts of multiplicity and plenitude which facilitated a conciliation between divine order and variety and between the divinely structured world and man's possibilities within it.[5] Thomas Starkey understood that "this sensible worlde, wherin is conteyned this wonderful varietie and nature of things is . . . a glasse of the divine magestie, wherby to man's judgement and capacitie is opened, the infinite power, and wonderfull wisdome of him."[6] Here, Starkey illustrated how important the intellectual-Platonic influence of the Florentine academy was for him, as it was for other English humanists such as More and John Colet.[7] But more interesting, he suggested the magus-Faustian strain that later Elizabethans would develop more fully to deal with the issues of order, immutability, variety, knowledge, and man's power.[8] For Starkey, stability and change, order and infinity, together, described the perfect world of God's "goodness" and man's possibilities. Variety, which was the source of man's power and accomplishments, signified God's gift to man and to the world. So, in the heavens, there were "infinite nombre [of stars] . . . ever keeping their certayn course and moving without al instabilitie," and on earth man's possibilities were numerous, but each was granted by God.[9]

By the 1560s during which Thomas Smith wrote *De Republica Anglorum*, England had witnessed four modifications in government and religious law in less than twenty years. Equally as disorienting and as socially dislocating had been a mid-century economic crisis whose ramifications England still felt. That Smith joined with many others in contemplating change and variety, then, was natural and understandable. That he offered temporal and secular explanations for economic, social, and civil problems was radical and innovative.

Although he posited the traditional Tudor commonplaces about governments, rule, order, and degree, Smith conspicuously refused to explain the social world in terms of religion or of natural law. Scriptural quotation served as illustration rather than as justification.[10] Smith observed that man's nature was cyclical and correspondingly commonwealths underwent change, too. The temporal end of all, he argued, was decay and degeneration. Nevertheless, he cautioned against useless change: "It is always a doubtful and hazardous matter to meddle with the changing of the laws and government, or to disobey the orders of the rule or government, which a man doth find already established."[11] Smith offered no biblical explanation for this, nor did he justify his view by arguing natural law.

As did other mid-Tudor and Elizabethan writers, Smith concerned himself with mutability and change. But unlike others such as Elyot, Sackville, and Cavendish, Smith did not fear mutability, but rather confronted it positively. Of course he was not ready to describe a new social order that was defined by change and variety, but he began to acknowledge the ubiquity of variety and change, and the necessity and possibility for man to direct this variety. From the obvious evidence of social variety and change Smith concluded the possibility of social amelioration. He was sure that "never in all points one commonwealth doth agree with another, no nor long time any one commonwealth with itself."[12] Yet the decay and failure of things, which evidenced change, and variety which argued for possible alteration of things, suggested that man could interfere with things as they were and possibly direct change. "The world is subject to mutability, that it doth many times fail," argued Smith. Consequently, it followed that the prince must make new laws and appoint new nobility "to honor virtue where he doth find it . . . [and] to plant a new where the old faileth."[13]

Smith's arguments advanced the nominalist understanding of political theory that Machiavelli had expounded. Each society, each commonwealth was separate unto itself and must be appreciated and evaluated in relation to its own circumstances rather than in relation to some absolute model. Indeed, Smith explained that there was no absolute, divinely ordered type of commonwealth; what was absolute was that each commonwealth should be organized to fit the circumstances defined by the time and the nature of its citizens. And since times and people change, so too would governments and societies; "mutations and changes of fashions of government in commonwealths be natural, and do not always come of ambition or malice. And that according to the nature of the people, so the commonwealth is to it fit and proper."[14] Smith understood, then, that governments specifically, and societies generally, were temporal and secular entitites, dependent upon time and human psychology. Unlike Elyot before him and Hobbes after him, however, Smith refused to reduce human activity and human psychology to a basic and uniform nature. No political philosophy based upon a discernible human nature or upon the naturalness of change and variety was forthcoming from Smith. Instead he offered to politics theoretical justification for temporal and secular considerations, and thereby brought the theoretical side of politics into the venue of practical politics. Like Hobbes, Smith knew that political theory and political experience were manifestly one. Politicians would have to realize that chaos, destruction, and anarchy resulted from forcing a people to accept a government unfit for their time and nature.[15]

Smith's political theory emphasized that things were meaningful in themselves and that in the case of civil society, for instance, the efficient cause was not merely a tool for an eternal and absolute final cause. Elizabethans began to appreciate such a nominalist understanding of religion, economics, law, and the quality of life in general, as well as of politics, without self-consciously realizing that such an epistemological foundation for evaluation was inimical to the traditional idea of order.

William Fulbeck, a legal theorist, acknowledged the uneasy and transitional quality of turn-of-the-century life. Voicing a theme common to numerous poets and theorists of this time, Fulbeck admitted that "of all worldly things time is most puissant."[16] In view of the temporal quality of life, Fulbeck reassessed the traditional understanding of social and religious experience and relationships. The basic hierarchical structure of society, the division of demesnes, and the difference in degree and calling of men purposefully expressed the law of nations, he argued. Fulbeck did not claim that such ordered hierarchy reflected any divine order or plan for society. Rather, an ordered social structure addressed the uncertain and temporal nature of nations and commonwealths; its direct social purpose was to prevent confusion and to propagate peace.[17]

Fulbeck's comments about variety in religious expression, while partisan in spirit, evidenced an evaluative sensibility similar to that of Smith's. Variety was not necessarily evil and diversity did not of necessity incur disorder, he maintained.[18] With reasoning that confirmed the special nature and essential purpose of things-in-themselves, he suggested that force of arms used to persuade in religious matters violated the nature and purpose of religion. "That which is against the nature of a thing can not tend to the effecting or preserving of that thing, but to the destroying of it."[19] Very simply, Fulbeck refuted the traditional understanding of a defined and ordered world in which all things were interrelated in nature and purpose as a result of their mutual relation to a single divine final cause.

Although the consciousness of time, variety, and change led to new considerations about evaluating life in general, and society and government in particular, Elizabethans created no new system of values against which all action and experience could be judged. Yet their inquiries led to a reassessment of the concept of society and of the individual in society. More particularly, Englishmen began to concentrate their attention upon their own time, their own country, and their own selves. Concern for isolated issues of specific interest increased. Temporal and secular investigations not only challenged the traditional Christian element in the idea of order, but helped exorcise the haunting concept of the superiority of antiquity that Renaissance studies had fostered. The young Francis Bacon had only contempt for those who would inpugn present conditions. "We need not give place to the happiness either of ancestors or neighbors," he cheered. For if one considered the state, religion, law, justice, traffic, learning, industry, and population of England, "it will appear that, taking one part with another, the state of this nation was never more flourishing."[20] For Bacon, quality of life had much to do with living in this world.

The concern expressed about time, change, and variety in Elizabethan England included continued discussion of political commonplaces and of the relationship between private desires and the public good. Traditional views about the three types of governments, about the duties of subjects and rulers, about good counsel, and about natural law continued to be reemphasized. At the same time, however, new attitudes about the relationship between law and order, and about society and the commonwealth, became more pronounced. As the gap between received ideas and new interpretations of social reality widened, self-consciousness about private identity became inextricably related to the redefinition of social meaning.

In *De Republica Anglorum*, Sir Thomas Smith, ever the rationalist, maintained that the king ruled with the goodwill of all the people and governed, not by natural law, but by the laws of the commonwealth. It was, argued Smith, incumbent upon the ruler to "seek the profit of the people as much as his own."[21] Where Elyot had sanctioned only public good, Smith acknowledged the demands and the attractions of public and private needs. He did not argue that the repression of private desires and the projection of these desires onto the public realm formed the psychological substance of the secular sovereign state, as Hobbes would a few years later; but he intimated the willful and articulated nature of the commonwealth, and suggested the important part the individual took in defining it. "A commonwealth," he said, "is called a society or common doing of a multitude of free men collected together and united by common accord and covenants among themselves, for the conservation of themselves as well in peace as in war."[22]

Charles Merbury exhorted Queen Elizabeth to secure the succession in a pamphlet couched in terms of a defense of monarchy "as of the best common weale." Filled though it was with the necessary references to traditional political commonplaces—Merbury's pamphlet praised the good monarch who favored the public weal and lambasted the tyrant who sought only his own pleasure, for example—Merbury's work attempted a fresh consideration of politics and society. Sounding like an early version of Francis Bacon, Merbury promised that his pamphlet would be a civil discourse, based not on precepts of learning but on reasoning and experience.[23] A commonwealth, he explained, was "an order of government observed in a citie, or in a countrey, as touching the Magistrates that beare rule therin: especially concerning that Magistrate, which has highest authorite, and is the principal."[24] Such a realistic, political, and secular definition of a commonwealth prefigured the political writings of Bacon and Hobbes. Merbury's concentration on secular authority transcended the traditional understanding of social order; his reference to the ruler's authority as "Power full and perpetuall over all his subjects in generall, and over every one in particular" was certainly an early suggestion of the theory of secular sovereignty that developed during the first half of the seventeenth century.[25]

If Merbury's arguments augured future political theories, they probably resembled the divine right theories proffered in the first decades of the seventeenth century more than Hobbesian ideas. Indeed, Merbury's monarch remained subjected to God's laws as well as to civil laws and to customs. Yet the sovereign monarch had an outlet even here. Merbury, anticipating later arguments of various hues, concluded that if civil laws and customs contradicted divine law, the monarch could certainly disregard them.[26]

The understanding of the relationship between law and social order altered as views about the commonwealth and about man's personal desires changed. Thomas Wilson believed that an eternal moral law existed which "standeth for ever, and is not altered at any time but is receyved from tyme to tyme."[27] He knew, too, that judicial law, though not as binding as moral law, was necessary for commonwealths. Yet Wilson's logic led him to suspect that little relationship existed between eternal, fixed law and order and the particular social order necessary in man's daily world. For that reason, Wilson argued, ceremonial law existed; it secured order and it affirmed moral and judicial law although it was a separate particular entity. Ceremony fulfilled the need "that there be an ordre in al our doings."[28] Order, then, resulted from ceremonial law, and while order was sanctioned by moral law, it was neither fixed nor defined by it. Wilson never suggested that man was responsible for the choices that would result in order or disorder. Indeed, his logical postulations understood very little relationship between the essential nature of things and man's actions. Man was free or he was bound, and "Goodness" was manifested in the body or in the mind, but man's actions neither directly resulted in freedom or slavery, nor defined what was or was not good or virtuous.[29] But neither did Wilson argue that the eternal defined the particular. Consequently, though he argued that ceremonial law secured order, Wilson never clearly posited that ceremonial law defined order, and indeed his entire discussion of the nature of social order, while rhetorically traditional, was conceptually ambivalent.

Other theorists stressed the logical and mechanical, but nonessential, understanding of secular law that Wilson emphasized. Law was a tool; its purpose was to insure social order. As this simple theory developed and acquired more secular attributes, the assurance and maintenance of social order slowly became associated with the necessary repression of private, individual desires and aspirations. Even the typically traditional and religious sentiment in John Ponet's statement that "Lawes be made, that the wilfull self will of men should not rule" expressed the understanding that law was purposively repressive.[30] Richard Crompton, too, suggested that law and restraint were like means directed toward the end of social order. Crompton added an extra burden, however: the sense of necessary restraint. "By law," he exhorted, "common peace and quietness is maintayned, if there were no lawe there would be no order, if there were no order al things would be confounded."[31] Indeed, it was the law that directed men toward a life consonant with the qualities received in the traditional idea of order. Crompton knew that only law led men to peace and to live "in firme concorde and agreement, in a unitie of wyll and minde, and in sincere love and charitie."[32] And while the feeling here was still one of a "fallen" rather than a naturally dis-ordered human nature, the prescribed remedy was the already developing tool of secular sovereignty: secular law.

William Lambard addressed the subject of secular law in his *Eirenarcha*, a practical treatise that discussed England's justices of the peace. These ministers of law and order were not about procuring some fanciful sense of natural unity or universal unanimity, argued Lambard. He recognized the absurdity of such a goal. Rather their objective was more secular, temporal, particular, and individual: to "suppresse iniurious force and violence, mooved against the person, his goods, or possessions."[33] Lambard emphasized his general point by interpreting Edward III's writ which first organized the justices. "By this writ," he wrote, "it is manifestly declared, that the Peace which he meant, was not an uniting of mindes, but a restraining of handes."[34] Lambard's explanation was perhaps the most explicit Elizabethan expression of the relationship between law, order, and restraint, and of the secular matrix that this relationship defined.

This appreciation of the relationship between law and order resulted in an understanding both of the necessity of law and of the necessity to make laws. Fulbeck advocated that "the making and maintayning of lawes is necessary . . . without law, which I interpret to be an order established by authority neither house, nor city, nor nation, nor mankind, nor nature, nor world can be."[35] Certainly, Fulbeck accepted the traditional view, whereby God was the author of all law. Nonetheless, within the scope of this traditional acceptance he suggested some radical possibilities. Already he hinted at the sovereignlike quality essential in secular law. And when he analogized that "the will, counsel and decree of the Citie is contained in the lawes, as the bodie can doe nothing without the soule," he resembled Thomas Craig and even James I discussing the sovereignty of monarchy far more than he did a typical Elizabethan order theorist.[36]

Elizabethan writers never managed to articulate a fully radical and innovative theory of order. Traditional explanation as well as unsanctioned radical theory existed side by side in many discussions. Thomas Floyd used biblical explanation when he argued that laws were invented "to salve the decayed estate and frailtie of mans nature" which was originally just, honest and well ruled, but had "swarved."[37] In the same pamphlet Floyd noted that all men were born to society and that "this naturall inclination of societie in generall, is in its selfe rude and barbarous, unles it be governed by counsel and tempered by wisdome."[38] Ambivalence prevails here. What is natural? Is order natural? Does man's social condition require order, counsel, and law because it is natural or because it is unnatural, that is, because society is the price paid for the fall from grace? Floyd never asks these questions, yet they demand to be asked because his concerns hover all about them.

Yet Floyd was perhaps the earliest theorist to suggest that the *natural* state of man demanded to be ordered, tempered, and resisted. Although he often had recourse to the traditional idea of a fallen nature that must be resisted, he went beyond the traditional when he allowed that "wee ought somewhat to restraine our libertie, diminish our credit . . . for the safetie thereof."[39] Certainly the idea of natural liberty differed from the traditional Christian sense of free will attributed to the Fall and consequently to fallen man. Floyd did not merely relate social order, and by suggestion, law, to restraint, but he tied social order and law to the concept of self-restraint; and in doing so hinted at the possibility of self-responsibility for order and for social definition. Floyd suggested that it was man's liberty, his naturally rude social propensity, that threatened social order, and consequently, man's well-being.

To suggest that Floyd understood, as Hobbes would some years later, the psychological relationship between self-repression and social articulation is, of course, absurd. Indeed, Floyd's articulated purpose was to instruct readers how to insure a perfect, and traditionally understood, monarchically led, Tudor commonwealth. But the insecurity that motivated the need to reemphasize traditional commonplaces was often a fillip for the unself-conscious expression of the innovative as well. It would be wrong to concentrate on the traditional in such work and to overlook that which perhaps in an unconscious and certainly in an unrefined way, interrogates, criticizes, and even to an extent negates received idea. This is especially so if we understand that such negations become part of the cultural dynamic of meaning redefinition even as they clearly are *not* common opinion. So if Floyd's discussion of liberty and safety indicated later well-defined arguments, so too his remark that "every man in generall loveth law, yet they hate the execution thereof in particular" prefigured Hobbes' understanding of the political and psychological nature of man.[40] Floyd described his "perfect commonwealth" in traditional commonplace fashion, but he hinted that social-personal relations of a most radical nature defined and maintained order in the real commonwealth.

The understanding of the relationship between the individual and the public world underwent profound changes as conceptions of the commonwealth and of the self changed. Temporal considerations about social purpose and Renaissance values concerning social decay and historical cycles altered the traditional understanding whereby the social world reflected divine order and purpose. At the same time a developing consciousness of self and a growing self-consciousness about private identity failed to provide the individual with alternative social and emotional meaning. While the altered understanding of the social world and of man's place within it made the sources of traditional meaning and identity precarious, the repeated references to the ephemerality of self-acquired ends made it clear that the more self-conscious individual was not facilely emerging as his own redemptive vessel. Certainly a social world that was understood to be inherently decayed and temporally defined could no longer stand as a representative redemptive order. There can be no denying that for ever-larger sections of the articulate population, life was now experienced and identity now sought in a social world no longer perceived to be reflective of the transcendent order.

Late sixteenth-century English thought and expression concentrated much attention upon the relationship between the individual and society. Certainly a great deal said on the subject repeated the standard Tudor view in which the individual's social place and vocation determined his identity and secured his relation with the divine through social hierarchy and purpose. The constant harping on this theme, however, suggested protestation more than it did affirmation, and certainly observers of all types depicted a real world very unlike the world described in such theory. George Whetstone, for example, compared the commonwealth to a building,

> all which partes firmely united together, does strengthen one an other, and the corruption of the least, by the sufference of Time, will turne to the confusion of the greatest, and therefore by the Lawe of Nature, the meanest person, in his vocation is sworne to travell for the publike benefite of his Countrie.[41]

Yet even this seemingly traditional appeal raised the specter of social decay and temporality; Whetstone argued that the public-private relationship responded to natural decay, to "Time." The prolongation of the mortal rather than the obligation of the immortal: is this the foundation upon which Whetstone's ideas are raised? Does the confrontation with change, time, and variety demand a reevaluation of the temporal? "If everie man in his calling would truly doe the office of a good common wealthes man," continued Whetstone, "Envie and all evill, would bee easily withstood."[42] In this argument Whetstone suggests that individual effort can overcome results of time. The successful commonwealth is the prescription, here; the individual and the commonwealth work together toward security, and perhaps toward redemption.

Thomas Floyd sounded the often-heard call for good, public-oriented, sagacious counselors. Both counselors and magistrates should prefer "the publike profit, before their particular commoditie," he assured his readers, and "the happie state of the common wealth flourisheth no longer, than it retaineth Counsellours."[43] Floyd repeated the views of nearly all who discussed public-private relations. Private desires should always retreat when faced with public demands; indeed, the prevailing opinion maintained that private and public purposes were antithetical. Only a very few theorists recognized that private desires and public goals could be reconciled. Ratcliffe understood that public vocations relied upon competent private vocations if they were to be successful, and that public wealth could be measured by the soundness and riches of each private person.[44] William Fulbeck went as far as suggesting that in certain instances private profit and self-interest neither harmed nor aided public good.[45] Such a conclusion implied that certain personal actions were to be self-evaluated and that a single received system of valuation did not exist.

The emancipation of self-interest was not accomplished overnight. Self-conscious, open approbation of private interests and quests remained a rare and radical exception during the late sixteenth and early seventeenth centuries, reserved often times for asocial protagonists in plays by Marlowe and Shakespeare. The more often self-interest could be identified as beneficial to the public weal the more comfortable individuals became about expressing self-interests and the more often individuals experienced emotional security from private endeavor. Hobbes was the first theorist wholly to predicate social meaning upon self-interest, but during the Elizabethan years the growing acceptance of self-interest and the consequent appreciation by some of the possibilities of self-definition indicated a cultural challenge to the received idea of order and to the traditional understanding of identity.

Sir Thomas Smith acknowledged the potential of self-interest in the pamphlet recently attributed to him, *A Discourse of the Commonweal of This Realm of England.* The editor of this remarkable pamphlet which analyzed the mid-century economic crisis, Mary Dewar, claimed: "The Discourse . . . recognizes self-interest as a perfectly acceptable force which intelligent government action should direct toward the common good."[46]

Two of the most practical Elizabethans, William Cecil, Lord Burghley, and Francis Bacon, acknowledged the attributes and the power of self-definition in different ways. Elizabeth's principal minister prepared a set of commandments to advise his son. Concerned with the choice of a wife, child rearing, patronage, and advancement in public life, Burghley geared his advice to a young self-interested political man. Young Cecil was not to burden himself with idealized, fixed prescriptions for behavior. A wife should not be poor, no matter how generous, "for a man can buy nothing in the market with gentility," advised Burghley.[47] Unlike Sidney who was caught up in the dramatics of a revived imperial and chivalric dream, Burghley advised Cecil not to rear his sons to be soldiers. Burghley's precepts for public behavior had little to do with any appreciation for natural social order and hierarchy: rather they expressed an astute psychological awareness about human action and motivation, and self-consciously recognized a self-serving and a self-defined social order. Cecil learned to be humble and generous with superiors, familiar and respectful with equals, and humble and familiar with inferiors.

> The first prepares a way to advancement; the second makes thee known for a man well-bred; the third gains a good report which once gotten may be safely kept, for high humilities take such root in the minds of the multitude as they are more easilie won by unprofitable courtesies than churlish benefits.[48]

Francis Bacon's life was a document of public service and of self-interest.[49] Whether he was advising James I and Robert Cecil, or preparing his Great Instauration, Bacon sought the advancement of his country, the advancement of learning, and the advancement of himself. Bacon endeavored throughout his career to reconcile his appreciation of man's private capacities with his understanding of the necessity of a wealthy, supportive, and well-served state. As a young man he warned against the Puritan influence in England, not from a religious point of view, nor from the perspective of political uniformity, but because he suspected that such influence represented a negative rather than a positive experience. "Again, they carry not an equal hand in teaching the people their lawful liberty, as well as their restraints and prohibitions," he declared.[50] Even then Bacon must have felt adrift in a world without fixed and finite external definition and sensed the positive possibilities for internal definition that he later articulated. In 1592, in his *Praise of Knowledge*, he noted: "A man is but what he knoweth . . . the truth of being and the truth of knowing is all one."[51] Man could acquire knowledge and thereby acquire his being. Bacon stepped gingerly into the existential void.

Perhaps the most touching example in Elizabethan England of the burden of public definition and of the attempt to reconcile the personal with the public, the internal with the external, was Queen Elizabeth herself. Encumbered by the received idea of kingship as well as by the Tudor name and her father's memory, and hounded throughout her reign to transcend her personal interests and to name a successor, or to marry and produce an heir, Elizabeth fought valiantly to define herself and to serve England too. She succeeded, while Shakespeare's kings, Richard II and Henry IV, attempting the same task, did not. To her subjects she was Astraea or

Gloriana, a body filled by mythic and spiritual as well as by historical forces. They surrounded her and she loved them, yet they demanded an heir or a successor; they needed one, because she was "Empress" and the "Empire" was transcendent. But how did Elizabeth experience these pressures? Certainly she wallowed in the myth and ceremony that surrounded her rule; but she added her own mystery to it also. The daughter of a dynasty, she resisted destiny, denied England a Tudor heir and defined herself in so doing; and she served her country too. She favored a conservative approach to political and social issues but allowed amazingly radical expression in foreign policy and in religious freedom. Indeed, she rescued England from the past and, at the same time, prevented others from turning it into an "Empire" of mythical and redemptive qualities. This she accomplished by maintaining a positive private persona within the public queen.

In 1559, speaking to her first Parliament concerning its request that she marry, Elizabeth assured them:

> I will never in that matter conclude anything that shall be prejudicial to the realm, for the weal, good, and safety whereof I will shun to spend my life. And whomsoever my chance shall be to light upon, I trust he shall be as careful for the realm and you—I will not say as myself, because I cannot so certainly determine of any other; but at the least ways, by my good will and desire he shall be such as shall be as careful for the preservation of the realm and you as myself.[52]

This statement reveals that, already, the young Queen Elizabeth had fashioned a suitable and conciliatory, a soothing yet firm and political tone with which to address her parliaments. But further, it reveals that the young monarch knew that she was both queen and Elizabeth. Although she showed herself ready to surrender Elizabeth to the queen and to England, she definitely suggested a burdensome distinction between the public and the private identities she assumed. Elizabeth's awareness that she could only account for Elizabeth, not for the queen, nor for any other private or public figure, allowed her to create a relationship between her public identity and her private self that was self-defined. By imaginative analogy, England, during her reign, began to create for itself a positive temporal and secular identity that smoothed its transition from a Catholic European world into a world of sovereign, individual states.

Determined to secure her self during her life and for posterity, Elizabeth refused the specious security of dynastic and imperial immutability. As she declared: "And in the end, this shall be for me sufficient, that a marble stone shall declare that a Queen, having reigned such a time, lived and died a virgin."[53] Elizabeth's virginity did not represent the surrender of self to public identity and obligations, as many suggest. On the contrary, her virginity raised her above the received idea of kingship and gave to Elizabeth self-defined temporal security and eternal meaning. Elizabeth discarded her dynastic and imperial claim to the past and to the future; yet she defined for herself an identity that transcended her temporal existence. Through self-action (or inaction) Elizabeth made her life her Final as well as her Efficient Cause. She was the Virgin Queen. She defined herself more fully than she was defined by the idea of kingship. Yet at the center of her self-definition was the magic of the passive act, and no one filled the magical ceremonial role of monarch more royally than did Elizabeth. Reconciling the creative and the received, the active and the passive, Elizabeth balanced the private and the public, the internal and the external, in a uniquely appropriate, magical, dynamic, and theatrical fashion. Only the fantastic and sensitive imagination of Shakespeare could, in Prospero, create such a character: as perfect for his time as was Elizabeth.

The possibilities of private identity and self-defined meaning and security notwithstanding, the consciousness of time and change continued to haunt the late Elizabethan mind. Secular and private values offered only a temporary emotional sense of purpose to men in the throes of a reevaluation of the relationship between the particular and the universal, between life and redemption. Men still sought emotional security beyond themselves and beyond the riches of this world. Floyd repeated the traditional homily to constancy, but managed only to emphasize his concern and insecurity rather than to consider any concrete proposals for human behavior.

> The treasure that men gather in processe of time, may faile, friends may relent, hope may deceive, vaine glory may perish, but constancy may never be conquered. . . . Constancy is the blessing of nature . . . the end of misery.[54]

In the face of the mutability and the meaninglessness of things, Floyd reemphasized the standard abstract conceptions that derived from the idea of order.

Confronting, as did Marlowe, the disturbing proposition that if the end of life is death what meaning has life, Richard Crompton attempted to revitalize the value of traditional behavior while accepting the need to find meaning in temporal activity. In a pamphlet dedicated to Essex, Crompton praised martial pursuits; but more important, perhaps, he addressed the question of the relationship between fame, honor, and a purposeful and meaningful life. "There is no difference between the greatest person & the meanest man, when they are both dead, if there be no vertues or deeds of fame done by them, whereby to commend their name to posterity," argued Crompton. If fame secured "everlasting," certainly traditional redemptive forces had failed.[55]

Although Crompton appreciated the need for man to act on his own behalf in attaining purpose and permanence in life, he rejected the more private, secular, and evanescent of man's possible acquisitions as being unsuitable to the task. Rather than sanction the new secular pursuits of self-conscious and self-interested men, Crompton wished to revive the magic and the honor of chivalric values, though the military action against Spain that he advocated barely resembled the type of chivalric engagement that he imagined as a model. Still, time was master and "riches and beautie do vanish soone away, but vertues and deeds of fame are everlasting; which sith our lives are short & momentarie, we must by this means make perpetuall."[56] Yet while Crompton acknowledged that personal fame acquired through self-action was the key to meaning and permanence in life, he resisted emphasizing the activity and concentrated his evaluative discussions upon the goal of the activity which he understood to be the true barometer of purposiveness. Circumventing, then, the dilemma that Shakespeare's Coriolanus experienced, Crompton argued that "there is no service more commendable then that which is employed, nor death more honorable (touching any worldly thing) then that which is given for the defence or safe gard of the Common wealth."[57] By so doing, he attempted to reconcile private and public obligations in a fashion that permitted positive self-expression and that identified the state as the primary source for order and meaning in life. In this way, the burden of self-definition could be denied, or at least deferred.

Meaningful deeds included the traditional service to the prince, but Crompton also suggested new ways to assure fame which indicated secular values and were anomalous to traditional chivalric attitudes. One could fund the building of churches, colleges, schools, and hospitals for the poor. As Crompton rhymed:

> None only for himself, but for his
> country too
> Is borne, and bound For her, the best he
> can, to do.[58]

Continual concern about social meaning and about the relationship between man's actions and "felicity" characterized numerous discussions in late sixteenth- and early seventeenth-century England. Could man locate purpose in temporal existence, and, specifically, in actions of his own definition and expression? Indeed, the great dramatists dealt with the possibilities and the ramifications of man's quest for self-defined meaning in a world lacking well-functioning redemptive institutions. Sir Richard Barckley described men's endless and fruitless quest to gain what "they could never find" in *A Discourse for the Felicitie of Man*.[59] In the preface to the work he categorically stated that no "mortall man hath in himselfe power sufficient to attaine to felicitie."[60] Yet he discoursed for over six hundred pages about felicity; and thus he expressed in a most poignant way the dramatic tension involved in self-consciously attempting to affirm received values and assumptions.

Although man would not discover felicity, he seemed evermore bent upon trying to attain it. Barckley understood the frustration thus engendered in man and explained that "an erronious opinion and wrong estimation of things was the chiefe cause that hindered their attaining to the end of their desires."[61] In so doing Barckley highlighted the psychological quality of Elizabethan consciousness that constituted, at least in dramatic form, a tragic confrontation with the social as well as with the private world. Barckley described what Marlowe and Shakespeare dramatized and what, among others, Essex, James I, and Bacon, must have experienced: the inability to reconcile new questions, opportunities, and expectations with traditional answers and explanations, and with received assumptions in general. Barckley argued that wrong opinions and expectations resulted in infelicity, but as did the great dramatists, he suggested that tragedy resided not in infelicity, but in the necessary quest for felicity and in the consequent frustration. The tragic aspiration to transcend the limited assumptions and possibilities that characterized the traditional way of life, though fraught with psychic danger, as Shakespeare made clear, eventually led to a new conceptualization and ritualization of identity and social meaning, based upon a negotiated relationship between self and other. But it was a very private and difficult journey, and certainly other alternatives existed. Barckley described a more common, less intense manner of dealing with similar concerns. Although he recognized the possibilities and the claims of self-consciousness and self-interest, Barckley advised a retreat into the world of traditional values. This necessitated a reemphasis and a revitalization of established values and a minimizing of new expectations and questions. The strategy required, however, a self-conscious approach to the question of felicity that no matter how undefined it was, reflected ambivalence and represented a compromise stance.[62]

Anxiety characterized Barckley's discourse. One senses it more in the almost compulsive nature of the discussion than in the argument itself. Barckley, determined as he was to define felicity, appraised all pretenders to the throne. He knew what was not felicitous. Felicity did not consist in pleasures, he argued; not in riches, not in honor and glory, not in moral virtue, not in active virtue, nor in philosophical contemplation.[63] The need to define felicity obviously resulted from failed received assumptions about felicity and redemption. But just as clearly, the numerous goals and activities beginning to be identified as socially and personally felicitous must have motivated Barckley's work.

Constancy and contentment characterized the man blessed with felicity; he "desireth no more [and] he is free from all perturbations & unquietness of mind."[64] Barckley argued from this that there could be no felicity in this life because "there is no estate of this worldly life voyde of troubles and calamities."[65] In so doing he reemphasized the traditional separation between this-world and otherworld, but he failed to alleviate the concern about temporal meaning. Consequently, in this life man appreciated only relative and negatively oriented felicity. The less one experienced sorrow and trouble, the closer one approached peace. Men could not quest for, nor attain, felicity in this life by positive action, Barckley suggested. Diminished infelicity rather than felicity measured this-worldly happiness.

Barckley never forgot that man had fallen: "he that looketh always to live happilie, seemeth to bee ignorant of the one part of nature."[66] Nothing in this world could, therefore, offer constant felicity, he maintained. Yet while he admitted to man's fallen nature, he hinted that human misery was self-defined. And in so doing he resembled Christopher Marlowe as much as he did any traditional free-will Thomist. As well as did his fallen nature, man's questing after felicity led him to misrepresent the object sought as the felicity sought. Although Barckley did not describe a restless, motive, search for felicity as natural in man, as would Hobbes, or maintain that in searching for personal felicity man defined his own damnation, as did Marlowe, nonetheless he suggested that man could alter his goals and thereby resist infelicity. "But the paines men commonly take in getting riches, and the care in keeping them, and the sorrow for loosing them, maketh men rather unhappie than happie that possesse them."[67]

Barckley, then, advised men to condemn worldly things. His was a religious message, but it derived from the same awareness of human psychology and the same appraisal of self-definition that supported Marlowe's vision. Barckley preached contentment. He implored men to "learne therefore to content thy selfe with thine estate, and that which God giveth thee."[68] Despite acknowledging self-conscious motivation and the human need to attain self-defined purposiveness, Barckley denied that man's self, either in isolation or in close relation with social directives, could secure purpose and felicity in life. Unable consciously to accept the power of self-definition, Barckley reaffirmed traditional values in the face of a psychological reality he understood to negate them. Struggling with questions of evaluation, just as Shakespeare did in *Troilus and Cressida*, Barckley declared that value, attributed by man to temporal things, was specious; these things were valueless. Consquently, if man did not value things in themselves, infelicity would certainly abate.[69] Aware of the tension and conflict involved in the constant quest for felicity and in the struggle to evaluate human action, he refused to advocate either a tragic confrontational stance or a comedic compromising stance with which man could engage the social world of order. Barckley's only conscious strategy directed man to retreat from self and from his search for temporal felicity and to accept wholesale his place in the received order of things.

Within his discourse, then, Barckley defined and rejected the developing psychological self-consciousness that sought meaning and identity in self-defined articulation. In historical opposition to Marlowe's unrefined, and to Hobbes' well-defined, advocacy of the good in the quest, Barckley firmly resubstantiated that the good was in the end, the final cause. Denying man's incursion into the realm of evaluation, Barckley reminded his readers: "God that hath made all things good, and is most good, and goodness itself is the felicity or beatitude and *Summum bonum* of man."[70] Yet, throughout, plagued with the intensifying awareness of the gap between human behavior and received assumptions, and confronted by "a wonderfull metamorphosis in mens minds and manners,"[71] Barckley had no recourse but to call upon self-interest and self-conscious articulation, the attributes that he identified as contributing to the dilemma of his times, to redress the condition of the world in which "men were never more religious in words, and never more prophane in deeds . . . and men seem to hate pride, and yet fewe follow humilitie: all condemne dissoluteness, and yet who is continent?"[72] Characteristically unwilling throughout to accept the positive possibilities of self-evaluation, and unable to rationalize self-interested action, Barckley still suggested that only through self-conscious restraint could man live well. Paradoxically, yet dramatically, Barckley told his readers that felicity or infelicity was, indeed, theirs to determine. Man had to control his desires if he wished to approximate contentment; but most important, Barckley reminded man, "let matters rather follow thee, then thou follow them."[73]

Certainly in an age as creative and as expansive as Elizabethan England was, it became increasingly difficult to deny self-interest and the possibilities of self-evaluation. Men required theoretical justification for social and personal attitudes that remained outside the moral parameters of the received idea of order. At the same time, however, no one self-consciously attempted to redefine the premises for social meaning. And any asocial imperative which suggested that the isolated self was in-itself the means and the ends of identity and meaning, was destined to be psychologically unfulfilling.[74]

As much of the previous argument has shown, no matter how rhetorically theorists defended or emphasized traditional assumptions and imperatives, the sensitive among them could not help confronting the awareness of change, variety, time, and self-interest which altered their perception of the social world and of the relationship between private desires and public demands. Those who described a relationship between self and society that still appreciated the priority imperatives of public order but who attempted to close the anxiety-producing gap between received idea and perceived reality, did not out-of-hand resist the directives of liberated self-interest and the possibilities of articulated self-consciousness. Indeed, social and political programs at this time, which, albeit unself-consciously, incorporated couched acceptances of the possibilities and the liberties of self-expression, began to describe an alternative set of evaluative imperatives which prefigured a redefined private-public relationship based upon self-definition and public utility.

Two vastly different programs, one social and one political, advanced by two very different Elizabethans, the educator Richard Mulcaster and the Puritan polemicist Peter Wentworth, exemplify fresh and innovative means to reconcile self-expression and the acceptance of self-definition with public utility and priority. Neither Mulcaster nor Wentworth fashioned himself as a radical, and both unself-consciously made use of traditional correspondences and metaphorical means of expression. Nonetheless, each, in his own way, and for different reasons, successfully enunciated publicly oriented programs which signified private interests and the expansion of self-conscious articulation.

Mulcaster's educational writings openly challenged time, change, and variety with the possibility of positive reform. Without negating the developing consciousness of self-identity, Mulcaster shifted the focus of the burden of frustration, expressed by Barckley and in Elizabethan tragedy, from the personal consciousness of insecure men to the social world of public identity. Mulcaster advocated necessary change, and self-consciously expressed ideas for educational reform. But he expected criticism, too: "If such objectives were not invariably raised to all attempts to turn either from bad to good, or from good to better. . . ."[75] Yet his keen perception of social reality and his determination to face this reality prevailed. "I may hope that the desire to see things improved will not be accounted fanciful, unless by those who think themselves in health when they are sick unto death."[76]

Mulcaster refused to surrender the real to the ideal or to subsume the temporal under the eternal. Unlike many Renaissance educators, Mulcaster cast a discriminating eye upon the glorious ancients. He taught that because social and historical circumstances changed, imperatives for one time were not necessarily credible during another time. "It is not proof," he cautioned, "that because Plato praiseth something, because Aristotle approveth it . . . it is for us to use."[77] And specifically when outlining an educational program Mulcaster stressed a realistic appraisal of social and personal circumstances rather than a received idealized picture:

> I mean to proceed from such principles as our parents do actually build on, and as our children do rise by to that mediocrity which furnisheth out this world and not to that excellence which is fashioned for another.[78]

For Mulcaster, education cemented a useful and a positive relationship between the individual and society, which modified the ordered hierarchic structure defined by traditional commonplaces. To insure the best social balance, some rich and some poor should be educated, claimed Mulcaster. In this way the rich would sometimes, but the poor would not always, be needy.[79] Young gentlemen should learn and exercise as all other boys did, said Mulcaster, and "the commonwealth . . . must be prepared to give scope for ability, in whatever class it may be found."[80]

Mulcaster realized that while hierarchy still described the social structure, traditional justifications for commonplace social rankings were expiring. Wealth distinguished commoners and authority characterized gentlemen, Mulcaster assured his readers.[81] Hierarchic social status, so defined, derived more from contemporary circumstances than from the received idea of order. Commoners or gentlemen, then, could be virtuous or vicious, rich or poor, since these qualities were self-defined characteristics, not externally ordered social absolutes.[82]

Indeed, Mulcaster stressed self-responsibility and understood that social good depended upon properly directing the possibilities of self-conscious articulation. Against those, for example, who wished to ban music because it caused bad manners, Mulcaster argued that bad manners derived from man's behavior, not from music: "If thy manners be bad, or thy judgment corrupt, it is not music alone which thou dost abuse, nor canst thou clear thyself of the blame that belongs to thy character by casting it on Music. It is thou that hast abused her, and not she thee."[83] Mulcaster described a more autonomous role for the self while he still accepted the traditional idea of order and the necessary relationships it employed. Nevertheless, the social repercussions of liberated self-interest could be avoided. And while Mulcaster expected the same end to private-social relations as did traditional theorists, he understood that only man, through active self-expression, could accomplish this end. Education, then, served to direct men toward their specific ends, to "serve their country well in whatever position they may be placed."[84] In this way, education in school replaced homilies in church as the coercive socializing institution of Mulcaster's Elizabethan world. In school a child received active preparation for life's demands and began directing his capacities toward accomplishing the ends of a good citizen.[85]

Finally, however, education as knowledge, or as the means to attain knowledge, transcended the strict locus of social purposiveness. Mulcaster aspired after the knowledge of nature and of art more in the "Faustian" fashion of John Dee and of Francis Bacon than in the manner of Elyot and other Tudor humanists. Although education had specific social ends, the acquisition of knowledge involved man in a complex reality so that

> by delineating with the pencil, what object is there open to the eye, either brought forth by nature, or set forth by art, the knowledge and the use of which we cannot attain to? By the study of music besides the acquirement of a noble science, so definitely formed by arithmetical precept, so necessary a step to further knowledge, such a glass in which to behold both the beauty and concord and the blots of dissension, even in a body politic, how much help and pleasure our natural weakness receives for consolation, for hope, for courage![86]

The "Baconian" emphases on attaining all knowledge, natural and artistic, and on utilizing such knowledge, characterize Mulcaster's remarks here. Man's aspirations, the exercising of self, can be socially useful without being, of necessity, socially derived.[87]

Although Mulcaster acknowledged the traditional commonplace analogy between music and harmony as did Elyot earlier in the century, he went beyond the expected, and argued that the study of music could reflect the dissension as well as the concord present, even, in the "body politic." The world Mulcaster investigated revealed a mixed reality of order and disorder, of natural and artistic creation; and the acquisition of knowledge about this world enlightened man about the personal and the cosmological, as well as about the social, levels of existence. Education and social utility directed self-expression in a world that metaphorically still resembled the traditional "world picture," but which in reality negated the well-defined correspondences that sustained the metaphor.

At the opening of the Parliament in 1576 Peter Wentworth, destined to be a thorn in the royal side for the next twenty years, presented a speech that Sir John Neale has described as "a distillation in startling and prophetic terms of all the radical utterances and yearnings . . . in the debates" of the previous Elizabethan parliaments.[88] Again Elizabeth's loyal subjects demanded that she settle the succession and that she deal with her treacherous cousin Mary. Elizabeth had forbidden Parliament to discuss what she claimed to be royal prerogative. In the 1571 Parliament she had warned Parliament that it "should do well to meddle with no matters of state."[89] Yet here was Peter, picking up where Paul Wentworth had not left off, demanding "liberty" and "free speech" for the Commons, and reminding anyone who would listen that "there is nothing commodious, profitable, or any way beneficial for the Prince or State but faithful and loving subjects will offer it in this place."[90]

Wentworth construed the well-being of the state as dependent upon the exercising of "free speech" and "liberty" in the Commons. "There is nothing so necessary for the preservation of the Prince and State as free speech," he claimed, and continuing, he argued that "free speech" in Parliament was constitutionally above the crown's interference: "Free speech and conscience in this place are granted by a special law, as that without the which the Prince and State cannot be preserved or maintained."[91] Indeed, Wentworth asserted that Elizabeth had committed "dangerous faults to herself and the state" and had opposed "herself against her nobility and people."[92] Consequently, Parliament must be free to act, to protect both the state and the queen.

Clearly, Wentworth's concept of the state and of Parliament's relation to the state revealed a modified understanding of the traditional Tudor commonwealth. Neale noted that whereas, traditionally, free speech had been a "customary right, if a right at all," during the Elizabethan years, it became a formal privilege, petitioned for and granted.[93] Parliament began to define a more representative and legislative role for itself; and the idea of order, which traditionally had been reflected in a monarchically directed commonwealth, now began to be represented by a legislative constitution defined by Parliament or by the king in Parliament.

Neale has directed our attention to what was most changing in the relationship between monarch and Parliament during these years. Parliamentary petitions had now become "the actual texts of laws enforcable in law courts, once the royal assent had been given; and that assent had been reduced to an unqualified yes, and the veto to an unreliable weapon."[94] The Tudor commonwealth, which had, of late, evolved into the Tudor state and the Tudor commonwealth, was now on the threshold of becoming only the Tudor state. As Neale saw, Elizabeth was intent upon keeping the "state" and the "commonwealth" separate: "The crown alone controlling one sphere of action and the crown in Parliament another."[95]

Wentworth's program opposed Elizabeth's. He wanted to join what Elizabeth fought to keep separate. Wentworth's concept of the state included the understanding that Parliament (Commons) could introduce unsolicited bills.[96] Constitutional change, then, and a new emphasis on "freedom" and "liberty," which led to wider connotations for both these concepts, characterized the shift from the Tudor commonwealth to the nascent Tudor state; and legislative autonomy and "liberty" became two sides of the same coin. When Wentworth argued that "free speech" and "liberty" maintained the state, he defined a social order that depended upon legislative responsibility and consequently, upon self-expression. In the state the public-private relationship was as close as it had been in the commonwealth; however, now the private assumed responsibility for the union. Wentworth, by joining "free speech" and "conscience" (see quotation from note 92), suggested the private burden incumbent upon private man for secular well-being.

Yet Wentworth maintained a fully conservative posture and never self-consciously assumed a radical and private burden of self-responsibility. Even as he attempted to legislate to the queen he justified his demands in the traditional language of Tudor body politic correspondences: "As therefore you are our head, shew your self to have dutifull care and love to your bodie, and if you may help it . . . you leave it not headless, as a dead trunk."[97] From one who was really about wresting away the royal prerogative, such traditional rhetoric sounds like mere cant. But the value of the traditional understanding of the Tudor commonwealth resided in just this capacity to admit a radical alteration in the conceptual structure of social order without necessitating wholesale personal alienation or social revolution.

Wentworth exhorted Elizabeth that it is "your dutie to doe that . . . the settling of the Imperiall Crowne,"[98] but he reminded her, too, that all claims to the succession should "thoroughlie to be tried & examined," in Parliament.[99] In this manner he neatly tied his radical claims for parliamentary expression and his own articulate self-expression to the "constitution" and to the general public good. "Such is, or ought to be," he preached, "the faithfull love of everie true hearted Christian subject towards his soveraigne, that feare to offend him may not shy us from performing of a necessarie, profitable, and honorable service unto God, our Prince, & countrie."[100] Wentworth's religious zeal and his continual emphasis on "wishing the good peace & prosperity of this our native country of England" served successfully to diminish the self-consciousness and the self-defining nature of his attacks.[101] At the same time the social orientation of his program provided a developing, self-defined private-public relationship the opportunity to mature.

The haunting vision of disorder and of inevitable chaos fed Wentworth's furies. Elizabeth behaved against God's "wil & nature, to leave us your people wittingly & willingly at random, to the rage & furie of hell & hellhounds," he despaired.[102] Fear and insecurity characterized his writings, and he supplicated to Elizabeth "with bitter teares . . . that you leave them not [her subjects] . . . in the lamentable & miserable case, to lose their lives withall they have."[103] Yet it is unclear whether he believed that man's helplessness or man's wickedness would instigate the ineluctable chaos that surely would follow Elizabeth's death. Was such doom predetermined or could man be held responsible for the state of his world?

Whether as a result of his Puritan theology or because of his insight into the social nature of man,[104] Wentworth seemed to believe that confusion and motion, rather than order, better described man's "natural" propensities. Consequently, he knew that order had to be maintained; he realized that policy and order were, in reality, the defining and the defined qualities of social equilibrium. Both were regulatory and necessary; after Elizabeth "all the mischiefes that the micheevous wit of man can devise, will be practiced . . . without controlment. For all the bandes of all good order and Policie will then be broken asunder."[105] So legislative responsibility, now, became the necessary source of regulatory policy for the entire state; in essence, Parliament, as Wentworth understood it, became order.

Nonetheless, Wentworth emphasized that religious duty motivated such private expression and parliamentary initiative. He told Elizabeth that she owed her crown to God and that not to settle the succession was "a grevous sin in you."[106] In reality Wentworth was usurping royal prerogative for the Commons; he was transferring to the Commons sovereign power and consequently was negating both the traditional mixed commonwealth and the developing divine right concept of monarchy. But as was the case with so many other late Elizabethan theorists, Wentworth was unwilling to admit the radical character of his program and was unable to assume the responsibility for defining an altered conception of sovereignty and of social order.

In almost direct contravention to the thrust of his appeals, Wentworth argued that Parliament did not legislate creatively, but described what is by reason and right. "The wisest men of the Realme are chosen out and sent to the Parliaments, not to determine or establishe whatsoever they will, but to advise, dispute and discerne what in reason & conscience they ought & should determine and conclude."[107] Such conservative rhetoric gave the lie to the earlier demands that Wentworth had made for Parliament, but it justified the place of Parliament in the developing state without necessitating an immediate struggle for sovereignty. Unable yet to stand conscious to the radical and innovative quality of his self-defined legislative program, Wentworth sought to fit it into the received paradigm of social order. And like Mulcaster, and others, he facilitated the fit by rooting his political program in the larger question of public well-being.

Still his assertion that Parliament would not choose the successor from contestants, but rather would settle the title upon the "rightful heir," comes across more as rationalization than as explanation.[108] "Parliament cannot defeat the lawful successor," Wentworth pledged.[109] A new king would be chosen not

> because it seemed good to the Nobles and commons so to do, and so to advaunce him; but . . . because he is next and true heire and successor . . . it is his right that he and none else should be advanced . . . he is not heire or successor because the Parliament declareth him to be so, but because he is so.[110]

Wentworth, here, avoided distinguishing between conferring the succession and settling it. In so doing he maintained the traditional essence of the Tudor kingship, acted in the name of public security and religious duty, and shrouded the private and legislative motivation that determined such political conceptualizing.

Although Neale is correct to point to the radical tone of Wentworth's parliamentary activities, it is not for his radical comments alone that Wentworth is such an important figure in Elizabethan political history; rather the theoretical framework within which he carefully blended a traditional sense of order with a distinct appreciation of social contract theory captured, more fully than did his radical speeches, the changing yet cautious political consciousness that characterized the dynamic quality of English political history at the time. Wentworth threatened the traditional order of things by focusing his attention upon the voluntary surrendering of government or sovereignty by the self to an other. In so doing he may appear as an opponent of traditional order, as one of Greenleaf's "empiricists."[111] Yet Wentworth suggested much more than this. In the following long quotation, Wentworth conciliated contract theory sovereignty with socially defined order much in the manner that Hobbes would a few years later. Social order was preserved, but it was redefined, too. Still, Wentworth accomplished a typical Elizabethan compromise by denying the temporal and self-defining quality inherent in social order; consequently, his political theory could not satisfactorily replace the received Tudor idea of order. Yet the similarity to Hobbes' arguments can not be overlooked:

> If all the people of the whole Realme by common & voluntarie consent, for themselves and their posteritie, do transferre and surrender the government of themselves and their state into the hands of some chosen, to be governed by him and his heires for ever, according to such and such laws, as they shall agree uppon, or have alreadie established; that they can not in reason (if he be willing to preserve their lawes) think, that the power doth yet rest in themselves, of which by consent of all the people jointlie giving, and the Prince receiving, they had formerlie dispossessed themselves. . . . And by consequence this posterity thus dispossessed of the power and interest of bestowing the right, cannot make voyde the act of their ancestors, in whome the saide power and right was actuallie and reallie, to dispose of their government before they bestowed it. Neither can the act of the rest of the members without the head & against it be of that power & force that the joined act of the head and whole members together is.[112]

Carefully here, Wentworth articulated the legal, historical, and the reasonable defense of parliamentary responsibility and, then, of public well-being. Characteristically he looked to aged custom when he wished to establish an argument. This type of conservatively innovative approach marked so very much of both theoretical expression and behavior in late Elizabethan England.

If Elizabethan England was a time of conservative innovation, or of innovative conservatism, then Richard Hooker, the finest political theorist and prose stylist of the period, was the ultimate Elizabethan thinker. Enveloping a far wider range of social and personal topics than did Wentworth's or Mulcaster's writings, Hooker's works dramatized the changes in conceptual outlook that characterized the development of a self-conscious individual response to the received idea of order and also to radical Puritan expression, that, as Walzer has shown was, in some ways too, a response to the demise of the traditional. An analysis of *Of the Laws of Ecclesiastical Polity* reveals in much the same manner that reflection upon Shakespeare's canon does, that with Hooker as with Shakespeare innovation and conservatism were not nicely balanced opposites. Indeed, the dynamic change beginning in these years was not to be found in the simple dialectic of old and new, of conservative and radical, but rather it issued from the conflict between various shades of innovation, between different articulated responses to conceptual questions arising from the lack of a securing received order. And it is as witness to the beginnings of this dramatic conflict that Hooker's work enlightens.

To label Hooker as a "representative" Elizabethan, or as a thinker who best represented the character of Elizabethan cultural or intellectual history is risky and unnecessary. Yet given the propensity of scholars to make of Hooker a medievalist or a modern, it is useful to emphasize that he shared with his contemporaries innovative as well as traditional conceptualizations of social meaning and order;[113] and in terms of intellectual history that does not merely suggest that he was a transitional theorist, paving the way from a medieval outlook to a modern one, either. Hooker's writings suggest a distinctive political discourse, which Hobbes and Locke conceptualize more particularly. Similarities of theme and expression between Hooker and Hobbes indicate the development of this discourse during the early years of the seventeenth century, but should in no way diminish Hooker's importance as a political theorist. That Hooker's innovative theories rested upon landmarks of the past and that Hooker "backed into modernity," as one scholar has suggested,[114] does not evidence the conflict of old and new, or the transitional quality of his thought or his times, but rather again reveals the characteristically Elizabethan concern with meaning and order and the simultaneous ambivalence about self-consciously redefining both.

There should be no forgetting that *Of the Laws of Ecclesiastical Polity* is primarily a polemical book. Indeed, late sixteenth-century England witnessed an outpouring of Puritan, Catholic, and Anglican literature which addressed questions of politics and political philosophy as well as of theology.[115] And while much of this literature remained at the level of diatribe, some of it transcended propagandistic limitations and confronted issues that were of wider thematic and historical interest. Hooker's work was the greatest and most conspicuous of this literature.

Addressing the Puritans directly in the "Preface" to the *Laws*, Hooker introduced the ideas about law and authority and about the relationship of law to state and sovereignty that he would develop in the complete work.

> A law is the deed of the whole body politic, whereof if ye judge yourselves to be any part, then is the law even your deed also. . . . Laws that have been approved may be (no man doubteth) again repealed, and to that end also disputed against, by the authors thereof themselves. But this is when the whole doth deliberate what laws each part shall observe, and not when a part refuseth the laws which the whole hath orderly agreed upon.[116]

Laws are neither immutable and eternal, or sovereign. They are subject to the deliberation and the will of authority; and this authority is represented by laws which signify social contract. Laws prescribe social behavior for private citizens.

Hooker understood that "positive law," not being natural, did not universally bind man.[117] While natural law sustained positive law, just as God sanctioned man's social authority, public approbation was still necessary to make a law.[118] Hooker emphasized the relativity of laws and customs; he saw that laws had a history and that this was a "human history" or understandable mutability.[119] Puritans did not understand human history, Hooker maintained, and he defended "a dynamic principle in human law" against the final scriptural law of the Puritans and against the static, inviolable canon law of the Catholics, too.[120] "Lawmakers must have an eye to the place where, and to the men amongst whome; that one kind of laws cannot serve for all kinds of regiment," he argued.[121]

Hooker recognized and affirmed man's capacity to "make" law as well as to interpret law: "All laws human, which are made for the ordering of politic societies, be either such as established some duty whereunto all men by the law of reason did before stand bound; or else such as make that a duty now which before was none."[122] As Professor Ferguson points out, Hooker implied that lawmaking should be "creative" rather than "regulative."[123] And in so doing he recognized the creative relationship between man, law, and social order and the further relationship between order and the historical perspective of change.

Hooker claimed that positive laws were not ends unto themselves, but rather were the specific means defined for specific ends. Consequently they were subject to alteration or abrogation, but caution was necessary in assessing laws which the custom of ages had confirmed. "The nature of every law must be judged of by the end for which it was made," maintained Hooker.[124] And this prescription held for positive laws of the Church as well as those of the state. Indeed, a law of God's authority should be changed or repealed if the purpose for which the law was articulated was either accomplished or changed. Especially ceremonial and judicial laws could be altered, Hooker wrote, if they failed to function adequately, even though they may have been sanctioned by God.[125] Hooker's appreciation of the relativity of law, even of the positive law given to the Israelites by God, and his sense of historical perspective are clearly expressed in his response to the Puritan emphasis on strict scriptural laws. Commenting on the laws of Moses, Hooker declared:

> If therefore Almighty God in framing their laws had an eye unto the nature of that people, and to the country where they were to dwell . . . then seeing that nations are not all alike, surely the giving of one kind of positive laws unto one only people, without any liberty to alter them, is but a slender proof, that therefore one kind should in like sort be given to serve everlastingly for all.[126]

The understanding that historical circumstance and change demand various types of law is evident throughout the *Laws*. Implicit in all Hooker says about positive law is the assumption that social order is dependent upon acceptance of secular authority; and operable laws are the "deeds" that signify the social contract between authority and the citizenry. All things good, and of God, and conducive to order could not be set down in the Scriptures. Church customs require obedience just as prescribed laws do, and common secular laws, too, necessarily fostered obedience, well-being, and order. The balanced relationship between law and order rests upon acceptance of authority and obedience. Therefore, Hooker, while he defended alteration of bad laws, cautioned against the wholesale abrogation of law or the disregard for ceremonies and aged customs. Authority must be earned, and maintaining of laws that have become customary appreciates the overall effectiveness of social coercion and influences social order. That men must be "induced" to obey law, Hooker accepted. The unwise alteration of customs or common laws, then, "must needs with the common sort impair and weaken the force of those grounds, whereby all laws are made effectual."[127]

Clearly, the conception that "society or government was founded on some kind of agreement or compact was coming to be held by a growing number of political theorists in the second half of the sixteenth century.[128] Sir Thomas Smith and Peter Wentworth developed ideas along these lines; so did Jesuit polemicists such as William Allen and "Doleman" (Robert Parsons). F. J. Shirley maintains, however, that Hooker was "the first Englishman systematically to formulate the idea of a social compact as the historical and legal basis of the State."[129]

For Hooker "the Laws of well-doing are the dictates of right Reason," and because of this, he argues, men naturally incline to sociableness.[130] They bind themselves as a society under the law of a commonwealth, thereby defining "an order expressly or secretly agreed upon touching the manner of their union in living together."[131] This order, of course, is dependent upon authority and obedience. In explaining the origins of government and of authority, Hooker described a relationship between the private individual and the public order that resembled, in certain instances, the one defined by Hobbes a few years later. Hooker declared that

> it seemeth almost out of doubt and controversy, that every independent multitude, before any certain form of regiment established, hath under God's supreme authority, full dominion over itself, even as a man not tied with the bond of subjection as yet unto any other, hath over himself the like power.[132]

While God was the ultimate source of authority and creation, man must assume the creative responsibility for political definition and order. "God creating mankind," Hooker continued, "did endue it naturally with full power to guide itself, in what kind of societies soever it should choose to live."[133] All legal government, then, was divinely sanctioned, and although man could define whatever type of government he chose to, Hooker still maintained that the primary cause of authority and, consequently, of order, was God.

Nevertheless, Hooker realized that man's behavior required that he contract social government. Without reducing political order to psychological foundations, Hooker noted that strife and a need for self-aggrandizement were common characteristics of pregovernment man. Because each individual acted in his own interests, conflict ensued, and Hooker explained, men agreed to end the conflict "by ordaining some kind of government public, and by yielding themselves subject thereunto."[134] Men, then, voluntarily submitted to an authority with whom they contracted, and in so doing they repressed their inclination to self-aggrandizement.[135] Without such an act which substituted a desired public end for a desired private one, "strifes and troubles would be endless."[136] Such a socially defined contract created an order that helped man fulfill his potentially natural sociability and order.

To facilitate this contract men "gave their common consent all to be ordered by some whome they should agree upon: without which consent there were no reason that one man should take upon him to be lord or judge over another."[137]

The implication here is that for practical purposes rulers are contracted guardians of the peace and do not rule naturally. Without refuting outright the traditional values of Tudor order theory Hooker granted that rulers were meant naturally to rule, but he concluded "nevertheless for manifestation of this their right, and men's more peaceable contentment on both sides, the assent of them who are to be governed seemeth necessary."[138] In so arguing he preserved the strength and the meaning that traditional values still furnished for late Elizabethan England. And though some suggest that his position reflected medieval election theory, and others seventeeth-century representative sovereignty, Hooker's careful confrontation with shifting social and personal values here and throughout the *Laws* expressed both the resilience and the transcendent meanings available to Elizabethans in the Tudor idea of order.

Hooker offered a commonplace explanation for the transference of power from the multitude to one particular person when he reiterated the accepted refrain that confusion reigns if "a multitude of equals dealeth."[139] But he suggested that the honor and the prerogatives that rulers assume were practical rather than natural gifts. Again addressing the relationship between private interest and public order, Hooker repeated that dissipation ensues when

> every man wholly seeketh his own particular (as we all would do, even with other men's hurt) . . . if for procurement of the common good of all men, by keeping several man in order, some were not armed with authority overall, and encouraged with prerogatives of honor to sustain the weighty burden of that charge.[140]

Although Hooker did not directly discuss the fact that men, who desire *privatum bonum* voluntarily organize government, the end of which is *bonum publicum*,[141] he realized that social order incurred a private burden of responsibility that could best be assumed through representative authority.

Hooker's political philosophy, then, maintained that authority and power derived primarily from God and that man was bound and subjected to this authority. At the same time Hooker suggested that social authority required the voluntary repression of self-interest, which enabled man to attain what human reason naturally understood as man's real "goal." Implicitly Hooker linked an awareness of the individual's responsibility for order with the consequently felt self-conscious burden of both authority and obedience. He only indicated that legislative representation would structure this relationship articulately and non-self-consciously. Subject himself to commonplace assumptions, he still argued that the English Parliament

> is that whereupon the very essence of all government within this kingdom doth depend: it is even the body of the whole realm; it consisteth of the king, and of all that within the land are subject to him: for they all are there present, either in person or by such as they voluntarily have derived their very personal right unto.[142]

Perhaps Hooker most subtly assessed the relationship between divine and human authority and the general limits on authority when he discussed church-state relations. Sharing the Erastian view of these relations with his Tudor monarchs, Hooker maintained that "the Church and the commonwealth . . . are

. . . one society, which society being termed a commonwealth as it liveth under whatsoever form of secular law and regiment."[143] The Church and the commonwealth include the same members, he said, and consequently "one and the same person may in both bear a principal sway."[144] Yet God does not appoint secular rulers with supreme ecclesiastical power. Such power "seemeth to stand altogether by human right, that unto Christian kings there is such dominion given."[145] God does, however, sanction the transference of authority from men to the king and in so doing authorizes the social compact and the laws and representative authority that must necessarily maintain it. Hooker recognized the special relationship thus defined, and he appreciated the limitations upon authority that characterized it.

> That the Christian world should be ordered by kingly regiment, the law of God doth not anywhere command; and yet the law of God doth give them right, which once exalted to the estate, to exact at the hands of their subjects general obedience in whatsoever affairs their power may serve to commend. So God doth ratify the works of that sovereign authority which kings have received by men.[146]

Sovereignty, then, was divinely ratified, but secular authority derived not from God, but from men. Hooker claimed that the king who secured authority and order was dependent on the polity: "Original influence of power from the body into the king, is cause of the king's dependency in power upon the body."[147] It was a dependency based upon authorship, not obedience.

Since positive law affirmed the social contract, the measure of the limited secular authority was the positive law itself. Although the king received his authority from this law, he was bound to it, as well. In this way, the king's authority was also the source of his subjection. Through the contract the king gained the multitude's power, but also the multitude's self-repressive responsibility. "The good estate of a commonwealth," Hooker advised, depends upon fear and love: "fear in the highest governor himself: and love, in the subjects that live under him."[148] The sovereign must fear to alienate his subject's love; he has assumed the burden of responsibility for civil order. The polity is always potentially uncivil.

In both civil and religious affairs the king ruled because the "law doth permit." Positive laws limit the king's power temporally and spiritually, and "the king alone hath no power to do without consent of the lords and commons assembled in parliament."[149] In this general context Hooker's famous, oft-quoted refrain about the king, the state, and the law offers fresh insights about Elizabethan values:

> Happier that people whose law is their king . . . than that whose king is himself their law. Where the king doth guide the state, and the law the king, that commonwealth is like an harp or melodious instrument.[150]

Although the language here was of typically "Elizabethan World Picture" style, the meaning suggested more than the commonplace understanding of the traditional idea of order. Hooker's political theory transcended the static and hierarchical character evident in received values and appreciated the self-defining quality of the dynamic yet balanced relationship between authority and responsibility that secured order.

Richard Hooker had a lot to say about order in *Of the Laws of Ecclesiastical Polity*. Not only did his idea of order reflect his dynamic political philosophy and his appreciation of the relationship between authority and policy, but it suggested his profoundly concerned assessment of personal values and of man's understanding of God and of himself, too. Traditionally scholars have casually caricatured Hooker as the perfect "Tudor order theorist," obsessed with divine order and degree and fully entrenched in Elizabethan commonplace correspondences. His remarks about harmony, unity, order, and degree are quoted alongside the famous speech by Ulysses in *Troilus and Cressida* to reveal the traditional, unselfconsciously expressed values of the "Elizabethan World Picture."[151]

"Without order there is no living in public society, because the want thereof is the matter of confusion, whereupon division of necessity followeth, and out of division inevitable destruction," claimed Hooker in the eighth book of the *Laws*.[152] Order is necessary for *public* society and as such must be understood in relation to the social compact contracted through law between authority and polity. The lack of such order leads first to confusion, then to division, and finally to destruction. But Hooker did not clearly maintain that order was natural and that the lack of order led to unnatural conditions. Indeed, he argued that

> order can have no place in things, unless it be settled among the persons that shall by office be conversant about them. And if things or persons be ordered, this doth imply that they are distinguished by degrees. For order is a gradual disposition.[153]

Here Hooker makes it clear that order must be defined, and that in fact, order can be understood as properly designated authority. As a goal, as political efficacy, it is of primary value; as structure, signifying received and divine nature, it is of secondary importance.

Due degree and the linked "chain of being" were divinely authorized, Hooker explained. In this way, he treated the idea of order as he had treated the concept of authority. God defined nature, the essential form that order described, but men were responsible for the existence of order in the public world. Order "is the work of polity, and the proper instrument thereof in every degree is power; power being that ability which we have of ourselves, or received from others, for performence of any action."[154] Throughout the *Laws* Hooker struggled to keep separate the essential and the existential meaning and cause of things. His success in this case allowed him finally to ignore the implications that were present in his own arguments: that man's self-consciously definitive capacities were ultimately responsible for authority, polity, and order.

Ferguson noted that Hooker was an ambivalent theorist. He yearned for the absolute but he acknowledged change. Indeed, he "recognized the possibility of purposeful change," and he admitted that order was often maintained by profitable changes in political, social, or personal behavior.[155] Holding Roman political history up to Machiavellian scrutiny, Hooker deftly showed that government, law, and order were malleable and changeable entities, and that security demanded acknowledgment of this truth.[156] Such awareness required a constant concern with valuation as well as with change. If things change and are still meaningful, how does one evaluate meaning? Struggling with these questions as did Shakespeare in *Troilus and Cressida*, Hooker noted:

> For what is there which doth let but from contrary occassions contrary laws may grow, and each be reasoned and disputed for by such as are subject thereunto, during the time they are in force; and yet neither so opposite to other, but that both may laudably continue, as long as the ages which keep them do see no necessary cause which may draw them unto alteration.[157]

Here again Hooker approaches an appreciation of the meaning of a thing-in-itself; and only subtle argument in other places in the *Laws* maintains the desired distance between essential and existential valuation.

Hooker's ambivalence toward absolute value paralleled his concern with variety and unity. To help explain contrary desire within one man, for instance, Hooker described two operations of man's will: the natural or necessary, whereby the will desires whatsoever is good in itself; and the deliberative, whereby the will follows the judgment of reason and "embraces things as good." Hooker offered the following didactic illustration: "Thus in itself we desire health, physic only for health's sake."[158]

Armed with this psychology of man, Hooker attempted to explain Christ's seemingly opposite desires or insights at his crucifixion: "inexplicable passions of mind, desires abhoring what they embrace, and embracing what they abhor."[159] Hooker's perceptive assessment of the variety of essences possible even in Christ addressed the issues of non-unity and valuation in a fashion that resembled Shakespeare's vision of complementarity.[160] He saw "in the human will of Christ two actual desires, the one avoiding and the other accepting death . . . so the will about one and the same thing may in contrary respects have contrary inclinations and that without contrariety."[161] Here again, Hooker labored over the relationship between divine authority and human responsibility, between the certain and the mutable, between God's will and man's basic propensity to desire ends contrary to God's will and purpose. With keen insight into human psychology, Hooker concluded that "no assurance touching future victories can make present conflicts so sweet and easy but nature will shun and shrink from them, nature will desire ease and deliverance from oppressive burdens."[162] Despite the theological questions, the religious language, and the use of religious metaphor in these discussions, Hooker's sympathy for the human condition and his understanding assessment of practical human psychology overshadowed the more traditional concerns. Hooker's man lived in the same confusing world that Shakespeare's did; and at times Hooker's man, driven by a motive nature that desired ease and avoidance of responsibility, strikes a haunting resemblance to Hobbes' man as well.

Yet such a comparison is too facile. Hooker's concern with change, variety, and valuation included a careful consideration of human motivation and of meaning and purpose for life and behavior. In the last book of the *Laws*, Hooker defended the rights of kings to stand above the judgment of the Church: "First, as there could be in natural bodies no motion of anything, unless there were some which moveth all things and continueth unmovable, even so in politic societies there must be some unpunishable, or else no man shall suffer punishment."[163] It is evident here, as many have maintained, that Hooker still comfortably and profitably argues within the Thomistic-Aristotelian framework. By correspondence,

Hooker determined that the laws of physical motion governed political activity. And by extended analogy it would not be incorrect to assume that he believed these same laws guided human behavior, too. An external, prime mover was responsible for all motion. Without the benefit of the new physics which would overthrow both the Aristotelian dependency on an external mover and the modified "impetus theory of motion," Hooker's recourse to established explanations in his confrontation with human motivation and valuation is understandable.

This traditional understanding of motion and of valuation, however, does not wholly inform Hooker's appreciation of the relationship between the absolute and the mutable. It often seems that Hooker makes practical use of the traditional framework. He tries to turn his concerns into ones that can easily be relaxed by arguments familiar to Elizabethan discussion. But the reiterated confrontation with variety and change, in the personal world and in the public realm, for Hooker, as for other Elizabethans, highlights the fragility of old assumptions and received ideas.

As did Marlowe and Shakespeare, Hooker assessed the plausibility of self-motivation. But he needed more than motion and existence as explanation for life. Unsatisfied by the fixed assumptions of the past, and unresponsive to Puritan psychology, Hooker sought to transcend the tragic potential that the burden of self-conscious responsibility held for isolated man. Values existed that were safe from mutability. There was justice beyond law and the course of civil justice. Sensitive as he was to the changing consciousness of his times, Hooker still relied upon unsatisfactory old solutions to new and pressing questions. And nowhere was this more evident than in his passionate appraisal of the issue that so captured late Elizabethan, and especially Marlowe's attention: what is the end purpose of all we do?

Could the answer lie in the mutable existence that is human history? Was man's reason for living defined by his temporal existence? Marlowe attempted to locate a certain existential, temporal meaning for life because he was not satisfied to let death be a positive factor in his life. In so restricting man's universe, Marlowe blurred his own tragic vision. Hooker demanded an immutable final cause, external from, but directly related to, and responsible for, human action. Yet he struggled to maintain this separate relationship between the transcendental and the existential. His writings reveal an uncertainty about an essential, theological absolute and a haunting awareness of the existential void.

Again Hooker confronted these questions of valuation and meaning with Thomistic-Aristotelian logic and theology. He demanded a single final cause for human activity, and he found one in the received commonplace assumptions of his times. "We labor to eat, and we eat to live, and we live to do good, and the good which we do is as seed sown with reference to a future harvest," he began.[164] Obviously, human activity in itself is not meaningful, "therefore something there must be desired for itself simply and for no other," he concluded.[165] And this "something," man's desired goal, his absolute final cause, had to be the infinite goodness of God. Yet even here the thrust of Hooker's demands and the tone of the decidedly expected resolution he offered tend to express an equal appreciation for man's capacity to aspire beyond his temporal limits and for God's divinely defined order. Hooker exercised the traditional idea of order and stretched its configurations.

Human history and divine order shape the world. Man strives; but does his aspiring nature remove him from God's order? Hooker tried to prove that it did not. To do so he had to accept natural mutability and multiplicity in the world; and he had to strengthen the ineluctable bonds that tied variety to unity and change to order. "All things that are, are good," he maintained.[166] Everything is divinely defined; multiplicity and change are godly and are God's creations. Hooker used the Renaissance neoplatonic "Idea of Goodness" to explain human psychology.[167] Man naturally desires, and moves toward his purpose in life; "there is in all things an appetite or desire, whereby they incline to something which they *may be*," he argued.[168] Hooker celebrated aspiration and drive in man, but he denied him a transcendent or a defining nature of his own. Man struggles to *be*, to accomplish a predefined self and meaning. He is not *becoming*. Change, desire, striving are not in themselves the locus of purpose and definition as they are in Marlowe's and in Hobbes' worlds. For Hooker, divine order circumscribes the mutable world of man. But since mutability existed, one had to accept it; and Hooker attempted to direct change and man's aspiring nature toward social order and away from the isolation of self-transcendence and self-definition. In this way the *Laws* resembles More's *Utopia* and Shakespeare's festive comedies. Rebellion in the public world and alienation in the private world are staved off by accepting variety within change and by ordering motion toward *being*. Following Renaissance neoplatonic tradition Hooker states: "There is in all things an appetite or desire, whereby they incline to something which they may be; and when they are it, they shall be perfecter than now they are. All which perfections are contained under the general name Goodness."[169]

Man then is defined by a finite final cause, Goodness, which itself describes infinite possibility and potentiality. In desiring this final end, man desires to share God's infinite Goodness. In this sense the absolute purpose is simultaneously unity and variety; variety is the creative energy which unity unleashes. Hooker noted that "there can be no goodness desired which proceedeth not from God himself . . . all things in the world are said in some sort to seek the highest, and to covet more or less the participation of God himself."[170] Man naturally tends to participate in God, to share his Goodness and his creative capacity. Hooker offered a powerful defense of the aspiring, self-actualizing man, which was, in fact, founded upon traditional Thomistic and neoplatonic theology. In an attempt to order and to explain the flux that characterized human behavior, and to secure the bonds between this behavior and transcendental purpose, Hooker conjured an Elizabethan Christian character who took his place alongside Marlowe's Faustus and John Dee's magus.[171]

Hooker's understanding of man's relation to God affected his idea of order and his appreciation of man's relationship to, and responsibility for, order. God makes himself known in the world, his creation. Knowledge and existence are inseparable; being and doing are one. "That perfection which God is, giveth perfection to that he doth," noted Hooker.[172] God, then, is the creator-creation, and he is known by the order of existence. Creation is the actualization of creator, as Goodness is good fulfilling itself. The descriptive principal, the meaning of creation, is not found in the created being itself, but is found in the formal, structural order that defines the creation.

Order, then, defines existence structurally and purposefully as first cause "whereof itself must needs be author unto itself."[173] Order is self-defining; it creates and is created by itself. It possesses the balance of active and passive, of kinetic and potential: qualities that characterize the creator. Consequently, the creator and the creation are one, also; and following from this, God and man are, equally, united through order. Man, like God, is potentially self-defining and responsible for other-defining. And as Hooker announced, the final cause of God's action is his will to do.[174] Consequently, purpose and definition are directly related to willing and doing.

The dramatic and poetic sensitivity for and appreciation of human capacity that characterizes Hooker's work reorients the traditional casting of his commonplace assumptions. It is in this fashion that he simultaneously reiterates and redefines cultural meaning. Dramatically, he declared that men must "enjoy God, as an object . . . that although we be men, yet by being unto God united we live as it were the life of God."[175] Men, then, must participate in Goodness, and must define order by actualizing and willing their potential order and being.

The redefinition of meaning is always a negation of social cohesion and it is potentially isolating. In a fashion Hooker appears here as another Faustus, or even Coriolanus. Followed to the extreme, his argument maintains that man defines his own final cause, objectifies it, absorbs it, and becomes it. In a culture whose redemptive institutions are fragile at best, man begins to look inside for the way out. The subjective and the objective merge, as do man and God, as do man and power or knowledge or any other defined purpose. This is the immanentizing and secularizing of traditional mystical, religious order, and it is potentially a tragic vision. Politically, of course, it lends itself to a consciousness that Hobbes defined so acutely, wherein the objectified final cause is the self-defined sovereign state.

Yet as did Barckley, and Shakespeare in the last plays, Hooker avoided the final secularization of order and in so doing retreated from the final tragic burden of self-conscious definition. Goodness and perfection are ultimately unattainable in this life, he concluded.[176] But as was Shakespeare's, Hooker's retreat was a self-conscious, planned, and purposive one; the source as well as the reason for the retreat was a profound consciousness of self and a concomitant consciousness of other. The nominalist connotations and the isolating ramifications of such awareness are apparent and have been dramatized by Shakespeare. Hooker denied the final distinction of man from other and attempted to negate it by reaffirming the unity of God and man. Ironically, such an articulation of unity exacerbated the consciousness of self and increased the felt burden of responsibility for order. In attempting to preserve the traditionally ordered world Hooker implicitly abandoned the metaphorical structure (discourse) upon which that order was dependent. His more inclusive description constituted a new order and thus helped redefine cultural meaning.[177]

A careful reading of the *Laws* uncovers Hooker's "ultimate sense of individual self-worth."[178] Herein may rest Hooker's greatest contribution to the cultural history and psychology of early modern England. His faith in human reason was firm. He maintained that, in the absence of direct revelation, man could trust his own reason and follow his own will. And he argued in favor of human authority in scriptural exegesis. If men have made such inroads into human sciences, he suggested, why could they not make similar advances in divine affairs as well?[179] But he emphasized that man's reason and deliberating will supported him most in determining the good in human history, and he counseled that reason had to distinguish between what man could know absolutely and what he could know only probably.[180] "Reason is the direction of man's Will by discovering what is good," said Hooker.[181]

Hooker's psychology of man still appreciated reason and will as separate forces. And while the "Will is Appetite's controller," its purposive function "is that good which Reason doth lead us to seek."[182] "To choose is to will one thing before another. And to will is to bend our souls to the having or doing of that which they see to be good. Goodness is seen with the eye of the understanding. And the light of that eye is reason."[183] For Hooker, the source of reason was external to man. Right reason discerned Goodness; and man possessed the capacity to tap this source and to choose to approach Goodness. But man could motivate and exercise the natural faculty of reason, and "education and instruction are the means" to do so, claimed Hooker.[184] Hooker's "will," deliberating among appetites, and choosing instruction as a means to exercise "reason" and "understanding," possessed a self-consciousness which indicated that responsibility for good and for order may be determined by a self in which reason and will, goodness and deliberation, were one.

What of this celebration of human worth and this appreciation of self-responsibility? Hooker claimed that human history, fraught though it was with flux and mutability, was meaningful; it could be understood, appreciated, and ordered. He knew that man had to govern in accordance with historical process in order to insure order, and he believed that man could do so. Ferguson argues that Hooker was an early "modern" because of his "perception of the relativity of human concerns . . . to the changing circumstances of time and place and his acceptance of the role of government as the positive factor in this historical process."[185]

Finally, Hooker, the innovative conservative, illustrates the dynamic quality of late Elizabethan political consciousness and the positive capacity with which Elizabethans confronted the demise of the traditional idea of order and struggled to redefine cultural meaning. Ferguson appreciates Hooker, but he is amazed that an Elizabethan with such a "medieval background" could articulate such views about man and government. He concludes that Hooker saw things the way they would be seen later; he was "pragmatic," claimed Ferguson.[186] Ferguson is too willing to attach "Renaissance" ideas to Elizabethan history and consquently, views Hooker as somewhat prescient. F. J. Shirley is equally unable to appreciate Hooker as an Elizabethan. Moving "beyond question . . . towards the modern Sovereign State," Hooker is halfway between a parliamentary sovereigntist and a divine rightist, Shirley argues.[187] Hooker did not define sovereignty or the sovereign state. And although he was neither a parliamentary sovereigntist nor a divine rightist, he cannot meaninfully be "categorized" as halfway between them, either. Divine right theories and the idea of parliamentary sovereignty developed separately as innovative alternative conceptual responses to the breakdown in the idea of order. While Hooker's views shared with both these seventeenth-century articulations this responsive character, Hooker never systematically confronted mutability and change with a fully formulated political philosophy. Though polemical in structure, Hooker's *Laws* extended beyond the boundaries of political order and described the historical and psychological premises upon which cultural meaning resides.

In 1594 the Jesuit Parsons published, under the name R. Doleman, *A Conference about the Next Succession to the Crown of England.* Parsons summarized his thesis in the preface to the pamphlet:

> It is not enough for a man to be next only in bloud, thereby to pretend a Crown, but that other circumstances also must concur; which if they want, the bare propinquity or ancestry of bloud may justly be rejected; . . . and this by all Law both divine and human, and by all reason, conscience, and custom of all Christian Nations.[188]

Parsons advocated the "election" of the next English monarch, and indeed, the "election" of a Catholic monarch. Reaction to the "Doleman" piece was vigorous and with responses from supporters of James Stuart, especially, the "Doleman controversy" ensued.

The Doleman controversy reflected conceptual changes in political consciousness that again confronted flux and mutability as well as concrete political-religious concerns that directed attention toward the succession question and ultimately toward the larger issue of sovereignty. Parsons' very polemical essay, though representing Catholic opinion, outlined, in an extreme and sometimes radical fashion, a theoretical position similar to those articulated by Smith, Hooker, and even Wentworth. At its most innovative and radical the pamphlet indicated ideas and theories that parliamentary sovereigntists would assume throughout the seventeenth century. Indeed history, probably with a twinge of irony, witnessed the resuscitation of this pamphlet by the supporters of Puritans during the Civil War and again by Protestant advocates during the Exclusion Crisis.[189]

The anti-Doleman pieces, most notably those by Craig and Hayward, were similarly propagandistic.[190] But they too suggested a soon-to-be-more-clearly-formulated political philosophy of sovereignty: the idea of divine right kingship. And even as foreshadowings of the more fully articulated theory, they indicated that divine right kingship was a positive and innovative response to insecurity and to the breakdown of the idea of authority and order rather than merely a conservative rehashing of the traditional idea of order.

Concerned with the issues of succession and sovereignty and dealing in both theory and political polemic, the Doleman controversy neatly and historically linked late Elizabethan and early Stuart political consciousness. In so doing it limned in bold strokes the beginning of the transition in political philosophy from Tudor order theory to seventeenth-century ideas of sovereignty; and it simultaneously suggested the tangible relationship between this philosophical transition and both consciousness and historical praxis.

Doleman's thesis rested upon the premise that the monarch was the people's elect and that he enjoyed his position by positive law. Blood lineage alone was not enough to warrant crowning a monarch. Other "circumstances also must concur": for Doleman, of course, first on this list of other circumstances was religion and service to God. The "want of Religion," which must be read as the want of Catholicism, is firm ground for excluding an heir apparent from the succession.[191] In fact, the monarch's highest duty is to guide his subjects in the way of the Catholic church and "whatever Prince or Magistrate doth not attend with care to assist and help his subjects to this end, omitteth the first and principal part of his charge . . . and consequently is not fit for that Charge and Dignity."[192] From this Doleman concluded that monarchs may rightfully be dethroned. And although the stated reason is "want of Religion," the justifying premise remained that the monarch has not fulfilled his duty to God and to his subjects. Doleman's emphasis upon the one Truth that the Catholic church represented suggests some interesting insights into the relationship between religion, politics, and psychology during this time. Reason teaches, affirmed Doleman, that "there is but one only Religion, that can be found among Christians."[193] Here, and throughout the piece, "one" is equally as stressed as is Catholicism. The emotional and psychological foundation that strengthened reformation Catholicism made it a particularly pesonal and historical phenomenon much as was Puritanism. Doleman declared that "unto me and my Conscience he which in any point believeth otherwise than I do, and standeth willfully in the same, is an *Infidel*, for that he believeth not that which in my Faith and conscience is the only and sole Truth."[194] This type of conviction offered psychological sustenance during a time of widespread doubt about authority, order, and unity. But even more than that, as for the Puritans, such personal conviction was a vital fillip to practical political action and for the formulation of a secular political philosophy.[195]

The king's power or authority is "a power delegate, or a power by Commission from the Commonwealth."[196] It is a restricted and limited authority, Doleman argued in attempting to establish a basis for his primary premises. Not only power and authority but laws of succession, too, derive from the commonwealth, and thereby "succession to Government by nearness of Blood is not by Law of Nature and Divine, but only by Humane and Positive Laws of every particular Commonwealth, and consequently may upon just causes be altered by the same."[197]

Doleman reiterated a theme common to late Elizabethan theorists: government and the superiority of magistrates derive from nature because men naturally are given to living together; yet particular forms of government are not defined by God but are "left unto every Nation or Countrey to chuse."[198] He then carefully analyzed the different forms of particular governments that had existed throughout history and the various types that existed in Europe in his day. What can be chosen by man, can be changed by man, Doleman said; and his historical aside proved that, indeed, man had altered his laws and customs throughout history. Although he recognized the larger defining structure of the Catholic church, Doleman appreciated that man was responsible for his own political and social order.

Although free to choose any form of government, man should choose monarchy, Doleman declared, because "it hath most Commodities and least inconveniences."[199] Certainly such a practical justification for government differed completely from common order theory explanations. Doleman admitted that God blessed magistrates, and consequently men must honor them, but he reemphasized that governments and magistrates are "a thing created by man . . . by mans free choice."[200]

Law limits authority in all cases, Doleman argued. "When men could not obtain equal Justice at one mans hands, they invented Laws."[201] Laws direct princes and help them govern; "all Commonwealths have prescribed Laws unto their Princes" which keep the commonwealth in good order.[202] But although he iterated a close relationship between law, order, and authority, Doleman never clearly defined the sovereign source of limited authority. Rather he offered another version of the popular late Elizabethan contract theory of government.

Doleman's theory, however, focused upon the coronation oath. A good prince contracts or vows to rule the people and to be ruled by law, Doleman maintained.[203] The coronation represents "mutual and reciprocal Oathes between Princes and Subjects."[204] Prince and subject, then, were mutually responsible for authority and order and their reciprocal relationship was characteristically active and passive.

By turning the coronation ceremony into a concrete contract, Doleman placed the burden of responsibility for government and magistracy upon the assenting commonwealth; but without assigning sovereign responsibility he created a political and psychological tension similar to that dramatized in Shakespeare's *Coriolanus*. At the same time, he refuted the received idea of kingship and the newly articulated theories of divine right kingship. The people owed allegiance neither to the king nor to the idea of kingship, but, through the reciprocity involved in the contract, to the representative symbols of this contract: the oath and the crown. The meaning of such synecdochic representation was specious. Yet it served as the metaphorical foundation for a more fully formulated concept of legislative representation. This whole discussion, however, emphasized that the throne could be empty. "Heirs apparent are not true kings, until their Coronation how just soever their Title of Succession otherwise be, and though their Predecessors be dead."[205] The ramifications of this changing view of kingship would read like a political history of seventeenth-century England.

It was not Doleman's choice of issues, but his methodology and his conclusions that suggest his radical and innovative posture. The general thrust of his theories negated the specific divine purpose described by the Tudor idea of order, and many of his arguments refuted outright the freshly conceived divine right theories as well.

A first principle of civil law, Doleman noted, is the division of goods. From this, and by the fact that the law directs the prince, Doleman denied the claim that "all mens Goods, Bodies, and Lives are the Princes."[206] Since the commonwealth prescribed law to help the prince rule his subjects, the same commonwealth could "lawfully chastise" the king if he failed to administer their directives. Indeed, to punish kings, Doleman assured his readers, is allowed by divine and human law.[207] With radical purpose and with anatomical illogic, he concluded that "the whole Body is of more Authority than the only Head, and may cure the Head if it be out of tune, so may the weal publick cure or cut off their Heads if they infect the rest."[208]

Doleman looked to history and to contemporary comparative governments for evidence to support his arguments. He listed oaths and agreements from the past as examples of "elections." He reported on the manners and customs of contemporary laws regarding succession and oaths in other European nations. And from Polydore Vergil he borrowed illustrations from England's history which proved that the direct blood heir had often been bypassed in choosing a successor to the throne. Such an emphasis on "history" celebrated the human and the temporal rather than the divine and the eternal characteristics of existence and purpose. If human history defined itself, then neither the received idea of order nor divine right conceptualizations made sense.[209]

Finally, Doleman's essay attacked the very idea of "legitimacy." Legitimate claims to the throne, if they existed at all, were meaningless, he argued. History showed this, and an understanding of the human derivation of law explained it without doubt. Consequently, since no final, fixed laws of inheritance and succession existed, he maintained, any reasonable claim, even the Spanish one, was valid. The people of England could decide upon any claimant. Concern for religion and order, then, could reasonably affect their choice.

The marriage of new political theories and Catholic polemic helped radicalize English political consciousness. Of the inevitable responses to Doleman, Craig's and Hayward's were the most direct and the most conclusive. And in rebutting the Jesuit, they adumbrated a political theory which in its later and fuller formulations would be known as the divine right theory of kingship. J. W. Allen claims that Craig and Hayward were still entrenched in the basic Tudor conceptualizations and that neither of them ventured as far theoretically as did James VI and I.[210] Nevertheless they, too, radically altered traditional concepts in order to attack Doleman's position.

Craig admitted that his prime objective was to disprove Doleman's assertion that monarchy was "neither of God's appointment, nor from *Natural Law*."[211] Kingship did not originate in law, as Doleman claimed, but "is of a Divine Original, ordain'd of God, and that no other form of Government . . . is so conducive to the safety of any Nation, nor is so agreeable to the Laws of God and Nature."[212] Both Craig and Hayward maintained that "Monarchy proceeds from the Law of Nature" and that monarchy, consequently, was the only natural form of government.[213] As Hayward declared, "natural reason teacheth us, that all multitude beginneth from one . . . that from unitie all things doe proceede, and are again resolved into the same."[214] From these convictions Craig concluded that "order therefore ought inviolably to be kept which God Himself instituted, *i.e.* that all other Creatures are to be ever Subject and obey man, men to honour their King, and the King to obey God to whom he is accountable."[215] Craig, then, used the traditional description of divine order to prove that kings received their authority directly from God.[216] And, without doubt, then, he maintained that no human law could alter hereditary succession because "rights of blood are not subject to any Civil bond."[217]

Besides these "natural" reasons for confronting Doleman, Craig and Hayward advanced practical reasons, too. Both feared the insecurity and the disorder that unfixed authority and "elected" kingship promised. Craig valued immutability and advised that "all alterations, especially in the Body Politick are to be avoided."[218] Hayward preached the immutability of nature, but he was conscious of the difference between ideal nature and manifest existence. Nature was not evenly received, he knew; it was immutable in *abstract* only.[219] Alteration of divine hereditary succession, he believed, would open the doors to disorder "wherein ambition and insolence may range at large."[220]

Craig attached his political theories to his interests on behalf of his country. He immediately established that public service was his motive because, he argued, his country was in danger.[221] His entire goal, he pleaded, is "the point of *Right* or Law, and that I have nothing before my Eyes, but the true Interest of *Britain*."[222] His emphasis on the public weal rather than on the religious obligations of king and subject conspicuously dramatized his opposition to Doleman. Even more significant, however, he emphasized the individual, temporal Britain, rather than the Catholic, eternal, empire that Doleman did. While Doleman used history and human law to validate eternal catholic truths, Craig established divine and natural laws to sustain the people and the state of Britain. He implored Elizabeth to secure the succession and realized that "the Estates and Substance of Private men can never be safe when the Publick is in Confusion."[223] The relationship between private and public interested Doleman only as it concerned questions of religious conscience. Craig's response, however, established a political theory, which included a self-conscious assumption of the necessary conciliation of private and public and which recognized the separate character of each. As Craig said, "Britain is the Common Country of both *English* and *Scots*."[224]

In refuting Doleman's position on the relationship of law, authority, and the succession, both Craig and Hayward continually pressed the issue of security. Election necessitates choice and variety, Craig claimed, but hereditary succession assures longevity and continuity. Besides, he commonsensically suggested, people are more inclined to obey a government to which they are accustomed than one to which they are not.[225] The means to security and order, he understood, were security and order. Hayward, too, seemed haunted by the specter of chaos. Kings, who originally governed by the law of nature and their own wills, prescribed laws in order to restrain people from exercising their authority and consequently, from changing governments.[226] Only his continued insistence on the natural origins of kingship prevented Hayward from admitting with Doleman and other Elizabethan theorists that government was, in fact, an artificial contrivance whose purpose was to restrain the natural impulses of people.

Since Doleman emphasized contracts and reciprocal oaths, Craig and Hayward, of course, opined about oaths and reciprocal responsibility, too. Craig explained that the coronation oath "is only a declaratory and overt act, with respect to the publick Right."[227] Oaths are not mutual or reciprocal, he noted, because "the King swears to God, and the People swear by God to the King," and, besides, he added, the king is king before the oath, anyway.[228] Hayward concluded, along these same lines, that oaths were voluntary rather than mandatory, because according to the common law of England—*Rex numquam moritus*—the king never dies.[229]

As far as reciprocal responsibility for authority and order is concerned, Craig denied that the people retained as a commonwealth the authority that each one surrendered as an individual by consenting to government.[230] While Doleman was unwilling to indicate a final sovereign source of law and government, Craig maintained that the king was above the positive law of the land and once government was contracted, he exercised full sovereignty. Certainly the commonwealth, which "is only a mute Body, and can do nothing of itself," could not assume sovereignty.[231] If at first, before positive law, the people elected a king to end the strife among themselves, he declared, "once they had chose him, they hereby transferred to him whom they elected and to his posterity, all their Power and Right."[232] In its understanding of authority and law and in its suggestion of sovereignty through representation, Craig's innovative, yet sketchy, political theory looked to Hobbes' concept of the sovereign state as well as to James I's divine right theory of kingship.

The emphasis on nature and on the law of nature in the anti-Doleman manifestos revived traditional explanations in order to justify radical political theory. As would other divine right theorists, Craig pointed to patriarchy, sanguinity, and heredity as the natural examples that supported his political positions.[233] But while he admitted that nature was responsible for inequality, he displayed little of the respect for due degree that Tudor theorists did. He defended Edward II's minion, Gaveston, whose indiscretions Craig listed as his "haughtiness" and his feelings of superiority to the nobility. Certainly he was guilty of stepping out of degree, and of overweening pride. Yet Craig asked, "which of the Peers did Gaveston wrong either by word or deed?"[234] It is interesting, then, to follow Craig's evaluation of behavior and action. Ironically, given his defense of Gaveston, he suggested that "we ought not to judge of the Justice of a cause, by the Success of it. . . . Tis the height of injustice to weigh actions."[235] Men's deeds are not the definitive ends of order and meaning. Responsibility and valuation, he maintained, transcend men's actions. Craig retreated from the self-responsibility that human history presented as a means to confront flux and the absence of fixed authority.

Hayward, too, sought order and authority outside of action and human history. He refuted the belief, which he attributed to Doleman, that a "fact is lawful because it is done."[236] As did Hooker, Hayward desired an essential order against which existence could be compared and judged. Doleman was wrong, he insisted, in maintaining that legitimacy did not exist. God determined that monarchy was the natural and perfect form of government.[237] Finally Hayward attacked Doleman's sources: history and the historian. Polydore Vergil was a suspect authority on allegiance and coronation oaths.[238] But so was history; "It is dangerous for men to be governed by examples," Hayward declared.[239] Precedents should not bind or necessarily instruct. Confronted with historical vicissitudes, and left without a definitive conceptual framework within which these changes could be understood, Hayward attempted to immobilize flux and to overcome insecurity by creating a conceptualized ideal, historical world replete with fixed political authority and naturally defined order.

CHAPTER 4

Refining and Defining: Jacobean England

And if *salus populi* be *suprema lex*, then though law and usage and prerogative were all against us, yet *bonum publicum* should be always preferred.

Francis Bacon, *Life and Letters*, vol. 3

Believing that I was born for the service of mankind, and regarding the core of the commonwealth as a kind of common property which like the air and the water belongs to everybody, I set myself to consider in what way mankind might be best served, and what service I was myself best fitted by nature to perform.

Francis Bacon, *Life and Letters*, vol. 3

Although James VI of Scotland peacefully succeeded to the English throne in 1603, English theorists continued to address many of the concerns and issues that had been related to the succession question. Certainly J. W. Allen was correct when he maintained that in the last years of the sixteenth century people "began to speculate about the origin of political authority and the nature of political obligation, about the question as to the ideally best form of government and the question as to where sovereignty lay in England and how much was involved in it."[1] And during the reign of James I of England, general concern about these issues and more specific discussion about the relationship between law, authority, and sovereignty and about the relationship between public, social values and private, personal desires increased. Different political theories advanced innovative concepts of order and authority and began to define a political and a conceptual relationship between order and secular sovereignty. Characteristic of much of the new political theory, as well as of essays concerning valuation and self-society relations, was a developing consciousness of self-responsibility and of the capacity for self-definition and a concomitant attempt to direct this consciousness of self toward the public good.

The basic premises upon which the divine right theory of kingship rested have been discussed many times and do not require a full elaboration here.[2] James VI and I, Edward Forset, and, a little later, Sir Robert Filmer most formally articulated the divine right theory during the first half of the seventeenth century.[3] Kings were divine because they "are called Gods" in the Scriptures, maintained James.[4] Indeed, he suggested to his Parliament in 1609, "if you will consider the Attributes to God, you shall see how they agree in the person of a King."[5] Additionally, he noted that by correspondence the king ruled the country as the father ruled his family, and by natural analogy, which both Forset and Filmer refined, he viewed the king as head of the body politic and the microcosm.[6] From this, of course, he argued that kings are accountable to God alone.

Figgis believed that divine right theory "was surely the product far more of practical necessity than of intellectual activity."[7] Devised as it was to refute Papistry and Presbyterianism, both of which viewed the state as a handmaid to religion, he argued, "not until the danger was past of a relapse into Popery and Presbyterianism, can the notion of Divine Right be said to have accomplished its work."[8] Theories of royal sovereignty, then, grew only in response to "extreme theories of popular rights."[9]

Divine right theory, however, never fully articulated any concept of sovereignty and certainly it preceded "extreme theories of popular rights" in the intellectual history of English political thought. And while there can be no denying that practical necessity and polemical considerations motivated much of divine right theory, it might be more useful to understand this concept as an innovative response to the breakdown in the traditional idea of order, and, consequently, related in terms of motivation to theories of popular and legislative responsibility and sovereignty, rather than as merely a conservative response to radical theories. To imply, as Greenleaf does, that divine right theory was basically renewed and revitalized Tudor order theory is misleading. Such a thesis ignores the historical and psychological character of this theory and overlooks the discursive metamorphosis that it illustrates. A discussion of the works of James I and of Edward Forset in light of these assumptions will focus upon the innovative qualities in their writings. It will highlight a continuing and developing redefinition of meaning which increasingly related a self-conscious reevaluation of self and society with a search for political order.

The second quotation that heads this chapter suggests the importance that Bacon placed upon personal service and duty to the commonwealth. Such service was twice satisfactory. It was self-expressive and it was socially beneficial. Social utility was the end that made self-expression acceptable. Any who experienced the urgings of private desires and of self-definition undertook the burden of demonstrating the socially utilitarian ends of his behavior if he expected social approbation. As Englishmen more self-consciously considered temporal ends, they sought means that would enable them to best serve society. In so doing, they became increasingly more comfortable with secular values, and identified political experience, pragmatic policy, and knowledge as worthwhile attributes to acquire. Similarly, a separation between civil and spiritual meaning for the social, political world could be made "only when utility [took] the place of divine law as the motive power of political policy."[10]

Obviously, experience, knowledge, and practical considerations have always been characteristics of good government and of good leaders. But when Machiavelli established them as the moral criteria for political behavior, he radically altered the history of political self-consciousness. Political order was a function of an independent set of moral postulates which, indeed, were identical with successful political behavior. In late Elizabethan and early Stuart England theorists continued to pay homage to an absolute morality which prescribed political ends and against which political behavior could be judged. At the same time, however, alongside this commonplace political understanding, they began to describe a new political world separate from the divine cosmos, self-defining and socially meaningful.

James I's political writings and speeches were often self-serving expostulations. But they expressed also a general theoretical position that was, too often, clearly discernible. In the introduction to his "Basilikon Doron," he emphasized the final authority of God's law. It was perfect and ultimately fixed and all laws derived from it, he maintained. Nonetheless, he pointed out immediately thereafter that "I only teach my Son, out of my own experience, what form of government is fittest for this kingdom."[11] Such a relating of experience, knowledge, and self-definition with truth, although not directly meant to contradict God's divine definition, conspicuously dramatized the more limited focus of James I's political and personal designs, and established the tone for his other works as well.

A little later in the same didactic essay, James advised Henry about the relationship between the Scriptures and efficient ruling. Again his practical and purposeful concerns dominated. Although the whole Bible is God's word, he coached his son to "delight most in reading such parts of the Scriptures, as may best serve for your instruction in your calling."[12] While one should not misconstrue the meaning of "calling" here by adhering to Weberian psychological sociology, it is clear that James understood that "calling" captured the dual sense of expression and of purpose. For James, "calling" revealed one's ordered and predefined place in the natural hierarchy, as it did for most Tudor theorists, but it also indicated the temporally purposive quality of and the utilitarian possibilities in self-expression.

James I's monarchy became as private as Elizabeth's had been public; and James found that he needed to demonstrate the public value of his behavior as tenaciously as did any other private citizen. In his opening speech to the Parliament of 1603 James noted that he always "kept Peace and amitie withall," and because of this "Townes flourish, the Merchants become rich, the Trade doeth increase, and the people of all sorts of the land enjoy free libertie to exersise themselves in their severall vocations without perill or disturbance."[13] He measured his value, then, by his ability to facilitate successful secular, temporal ends in the commonwealth. And much as the merchant pamphleteers married their gain to society's gain, James championed himself as the protector of the common wealth.[14] Parliamentary recalcitrance exacerbated James' sense of isolation and continually prompted him to reiterate his understanding of the pragmatic relationship between the king and the state: "For if the King want, the State Wants, and therefore the strengthening of the King is the preservation and the standing of the State; And woe be to him that divides the weale of the King from the weale of the Kingdome."[15] Practical considerations demanded that a radical definition of the monarch's place in society focused upon the secular, temporal wealth of society rather than upon the traditional order which described the Tudor "commonwealth."

James I's speeches and essays belabored the theme of the individual's relationship with and responsibility for society. He recognized the distinction between private values and public good; and while he accepted the worth of the private he sought to reconcile the two as much as possible. A king's ends are the public good, he maintained; and as a king, he advised Henry, he should publicly serve justice and equity by "establishing and executing . . . good Lawes among your people."[16] As a private man, he continued, you can "teach your people by your example."[17] But James I struggled with the king's two identities: the public role and the private man.[18] He allowed that the king's goal was finally the welfare and peace of his people. He subjected "his own private affections and appetites to the weale and standing of his subjects."[19] But he continually acknowledged that this subjection was a voluntary one and suggested that, in reality, a king shaped his public role privately. Indeed, the king's conscience determined his public behavior. And to actualize the public role of kingship, "kings . . . will be glad to bound themselves within the limit of their Lawes."[20] A king, then, was bound to the law only by his "good will"; "a good King, although hee be above the Law, will subject and frame his actions thereto, . . . of his owne free-will."[21]

James' understanding of the self-defined character of the king's behavior informed his explanation of social origins and of the relationship between the king, the state, and law. Although he argued from Scripture and by analogy that kings were divinely ordained and were judges "set by God over them [people]," he tried to establish by historical illustration that kings preceded laws and society.[22] Kings established society, he declared, "before any Parliaments were holden, or lawes made: and by them was the land distributed . . . states erected and decerned, and formes of government devised and established: And so it followes of necessitie, that the Kings were the authors and makers of the Lawes, and not the Lawes of the kings."[23] Such "willing" kings exercised an authority and a creativity that prescribed and articulated social order.

Before society, and before laws the, kings' "wills . . . served for Law; yet how soone Kingdomes began to be settled in civilitie and policie, then did Kings set down their mind by Lawes, which are properly made by Kings only."[24] Such a naive view of the origins of society cannot minimize the fact that James believed that societies and laws were the publicly created expressions of private, individual kings. Laws represented the king's definitive ordering of society. At any time the king could modify the social order and "make new lawes and statutes, adjoyning the penalties to the breakers thereof, which before the law was made, had beene no crime to the subject to have committed."[25] The king's actual authority for order, and his power to create law are glossed over by attributing the king's rights to divine ordination. Yet for James neither order or the king's social position are predefined. Rather, order is continually actualized through the king's creative expressions: his lawmaking and his governing. And the king's social place within this order is determined by his willingness to "rule my actions according to my Lawes."[26] In many ways James I expressed a Faust-like inclination to self-definition which was tempered by divine right rhetoric. And because of this his theories are important to the history of legal, secular sovereignty.

Early in the reign of James I of England, Edward Forset, a rather obscure justice of the peace, published *A Comparative Discourse of the Bodies Natural and Politique*. Forset's treatise was an articulate defense of the divine right theory, but it cast this theory in a fresh light by stressing the natural sanctions for political order. Although Figgis credited Filmer with transforming the analogy for political order from the theological to the natural realm,[27] it is clear that Forset's early attempt to base political theory upon natural analogy bore witness to the breakdown in the traditional idea of order, and was directed toward describing anew "some immutable basis for politics."[28]

Since it defended order and monarchical authority, Forset's pamphlet was, in many ways, a traditional political treatise. But from the very beginning Forset's analogies and allusions transcended the traditional concepts of order and of hierarchy and described a world in which man assumed the responsibility for creating political order. He fashioned his divine right theory and his concept of order from a different understanding of man's relationships with God and with nature than Tudor order theorists had.

In the introduction to his pamphlet, Forset compared the commonwealth to the whole world and to the body of a man. Using neoplatonic and Hermetic references and quoting Trismegistus, he described man as a giant stretching with his head and shoulders into the heavens while his feet firmly touched the earth.[29] Borrowing from the mysterious cosmology of Pico, Bruno, and John Dee, Forset established an analogy for political definition and order that transcended traditional hierarchies and celebrated man's place in both the natural and the political orders.

What is most striking in Forset's analogies, then, is that man has become the referent for the rest of creation. "In the very composure of man, there is manifestly discovered a summary abstract of absolute perfection, by the which as by an excellent Idea, or an exact rule, we may examine and exemplifie all other things."[30] Such a man surely resembled Prospero. The all-embracing order that defined the relationship between man and the cosmos still existed for Forset, but now man was emerging as, at least, the conductor if not the composer of this cosmic orchestration.

God's full, created universe was infinite and various; and man's chosen purpose was to interpret this creation, thereby discovering the ubiquitous natural relationships and analogies therein, and consequently, to determine order in multiplicity and variety: "It is the greatest miracle of God's powerfull wisdome, in the innumerable frames of things to make infinite variation; then it must needs be a great work of the wit of man, in such multiplicitie of difference to find out the well agreeing resemblances."[31] By combining the concept of natural order with the neoplatonic appreciation of man's responsibility for discovering order, Forset advanced a respectable and seemingly traditional theory of order which required the dynamic exercising of self-expression and of individual responsibility. Men created and structured their political world by imitating God's natural creation and by establishing a fixed order which resembled God's natural order. As Forset noted,

> the incomprehensible wisdome of God, in the composing & ordering of his works in nature has so dignified them withall perfection, as that they be left unto us as

eminent and exemplary patterns, as well for the consolidating, as for the beautifying of that we worke by arte and policie.[32]

In most instances Forset's political theories resembled James' and presented a strikingly conservative picture of the ordered, monarchical state. He did not, however, turn to theological argument or to scriptural allusion in advocating his theories. He based his concept of political order upon a modified neoplatonic natural philosophy that suited his particular concern with civil disorder. As in man there is an active ruling soul and a passive ruled body, so in the civil state there is an active ruling sovereign and passive ruled subjects, claimed Forset.[33]

Continuing his discussion of the "body politic," Forset noted that each state needed a good mixture of four "elements": the generous, the learned, the yeomen, and the "traffickers."[34] But it was necessary to establish the proper mixture, and thereby to create social order. Forset explained that "right needfull it is in any commonweale, to contrive the true and proportionable mixture of these four Elements, lest they be put at odds, reverting to the originall repugnances of their nature, they do fill the state with hatefull strifes, in the stead of blessful peace."[35] Order derived from a contrived blending of various elements and was dependent upon the "ordering" capacity of the civil rulers.[36] As God ordered the diversity in nature, so too must prudent policy order the diversity in civil society. And finally, as each body experienced growth, power, decay, and downfall, so too did each commonwealth.[37]

Forset based social utility upon natural analogy, too. He believed, as James did, that private behavior had to be directed toward the public good. And while he appreciated man's need to create order from diversity and to identify unity in chaos and mutability, he explained that each man must repress his private interests. As each action of the body is done for the benefit of the whole body, so too all is done for the common good, he argued. "It is not therefore called a Commonwealth, that all the wealth should bee common, but because the whole wealth, wit, power and goodnesse whatsoever, of every particular person, must be conferred and reduced to the common good."[38]

Unlike many of his contemporaries, Francis Bacon openly admitted the good counsel of Machiavelli: "we are much beholden to Machiavel and others, that write what men do and not what they ought to do."[39] Yet many heeded, with Bacon, Machiavelli's advice. During the early years of the seventeenth century, discussions about politics and political theory became increasingly more practical and more involved with what was rather than with what ought to be. Particularly, theorists stressed secular values and temporal ends when they argued about social responsibility and about private-public relations. Politicians and theorists debated about the true definitions of prerogative, sovereignty, and authority, but nearly everyone agreed that the public good was the ultimate end of all social and political activity and that practical policy was the means to that end.

Secular values became involved in the continued confrontation with time and change, too. More and more theorists admitted that the civil, historical world was not defined by any divine and immutable set of laws; and consequently they maintained that an awareness of changing circumstances led to better policy.

Bacon recognized that social relations and the idea of kingship were not immune from time's alterations. Yet he feared that King James refused to face this truth. As he wrote to Northumberland: "I told your Lordship once before, that (methought) his Majesty rather asked counsel of the time past than of the time to come."[40]

In the continuing examination of private identity, secular values acquired greater importance and acceptance. As men grew more self-conscious about their ability to shape the direction of their own lives, and as they began to experience meaning in personal endeavor as well as from social, collective identity, they began to evaluate taken-for-granted ethical foundations for behavior. Certainly, they continued to defend their behavior as beneficial to the public good. But such advocacy could not merely recall traditional arguments to bear witness to its case. Since the idea of the public was now becoming divorced from the strictures of any theological first cause and was becoming reformulated as an end unto itself, publicly directed private activity required ethical referents that reflected this redefinition. Bacon celebrated man's emancipation from predefined ends and advised man to choose "honest and good ends," ones that were "within his compass to attain," because, as he argued, man had the power to mold his character and personality.[41]

During Elizabeth's last years, questions about the succession led to an active consideration of the rights of Parliament and ultimately to an assessment of the relationship between the monarch's prerogative and the Parliament's liberties. Still, Tudor theorists agreed in general about the structure and the purpose of English institutions; and the traditional idea of social organization prevailed. Even when a new issue, monopolies, again highlighted the question of prerogative there was no reason to expect that the basic understanding of the idea of kingship and of social institutions, in general, would radically change. Bacon represented conservative opinion on this issue, and during the monopolies discussions in November 1601, he declared: "For the prerogative royal of the Prince, for my own part I ever allowed of it; and it is such as I hope I shall never see discussed."[42] He opposed a law which would inhibit the prerogative:

> The use hath been ever by petition to humble ourselves to her Majesty, and by petition desire to have our grievances redressed; especially when the remedy toucheth her so nigh in point of Prerogative . . . I say, and I say again, that we ought not to deal or judge or meddle with her Majesty's Prerogative.[43]

Restraint, however, did not prevail. Parliament challenged James' prerogative over domestic issues such as impositions and purveyance and over foreign affairs, too. Indeed, on February 27, 1610, the Commons objected to certain definitions published by John Cowell in *The Interpreter*. Among the objectionable terms were *King*, *Parliament*, *Prerogative*, and *Subsidy*; Cowell's definitions represented James' personal understanding of kingship and of prerogative. Parliament subsequently began openly to define these terms also, and on March 8, James replied to the Commons that although "it was dangerous to submit the power of a king to definition . . . he did acknowledge that he had no power to make laws of himself, or to exact any subsidies *de jure* without the consent of his three estates."[44]

Beyond being a confrontation over specific issues, the differences expressed about the prerogative and subsidies, for example, represented an inquiry into the nature of authority and order and assumptions about where responsibility for government resided. The nature of the English kingship and the identity of the Commons had developed to such extents that both transcended the boundaries of social organization as defined by the Tudor idea of order. As Bacon knew: "The King's sovereignty and the Liberty of Parliament are as the two elements and principles of this estate."[45] Good politics demanded that these dynamic institutions respect each other; and political theory began to define the "idea of the state" in which these political institutions, by policy, tempered each other's authority, insured social order, and consequently facilitated the public good.

In the first decades of the seventeenth century, order became more identified with political activity and the state and less associated with social obligation and divine purpose. Although he advocated the king's prerogative, Bacon understood Parliament's growing importance and self-confidence. In 1612 after Cecil's death, he tried to coax James to call a Parliament. "But the great matter and the most instant for the present, is the consideration of a Parliament, for two effects: the one for the supply of your estate; the other for the better knitting of the hearts of your subjects unto your majesty."[46] During 1613 he continued to exhort James to call a Parliament and to act more familiarly with it. "I conceive the sequel of good or evil not so much to depend upon Parliament or not Parliament, as upon the course which the K. shall hold with his Parliament."[47] Raleigh, too, realized that the king's attitude toward Parliament, and through Parliament toward his subjects, was important in maintaining good order. He counseled that the king should use the "Purse" only with consent of Parliament, and "to use it, that it may seem rather an offer from his Subjects, than an Exaction by him."[48]

Bacon understood that Parliament was experiencing a growth in self-confidence and self-consciousness. While he rhetoricized about the ancient obligations of kingship and Parliament, he appreciated that both these institutions had changed. He sensed that the Commons' increased assumption of responsibility for the public welfare characterized this change and he hoped to direct this energy toward the benefit of society and still to keep it from confronting the king's prerogatives. He urged James with the kind of "order rhetoric" that might reach the pompous monarch: "Until your Majesty have tuned your instrument you will have no harmony. I . . . think it a thing inestimable to your Majesty's safety and service, that you once part with your Parliament with love and reverence."[49] But practically, he advised the king to give the Parliament more of the country's business to discuss; perhaps if it concerned itself with trade it would lose interest in the king's business.

From every side, and in all arguments and theories, taken-for-granted descriptions of society and social purpose gave way to simple assumptions about the well-being of the subjects. Theorists and politicians alike divorced the state from first causes and principles; of primary importance was not why a specific government or state existed, but rather, what it provided.[50] Certainly everyone cited historical custom and tradition, and everyone attempted to prove that he represented the long established way. But, again, this can be understood as necessary and honest rationalization. Raleigh used "order rhetoric" often, describing society by recourse to correspondence theory. But when he discussed politics he used few religious illustrations and rarely spoke about natural law. Even when he discussed the best form of kingship, he gave in to practical considerations. Elective princes are more "Ancient," he claimed; and kingship by "Succession is better," but, regardless, he concluded, the "chief and only Endeavour of every good Prince, ought to be the Commodity and Security of the Subjects."[51]

In 1628, Sir Henry Marten told members of both houses, "We are assured, that there is no *Soveraigne power* wherewith his majestie is trusted either by God or Man, but only that which is for the protection, safetie, & happinesse of his people."[52] And Henry Wright realized that "every forme of government is so subject to change [and] alteration," that one had to search to find the best forms for the welfare of "Kingdome and Country."[53] Indeed, Wright denied that any government "bee absolute and perfect," since each form of government "doth chiefly depend upon the nature and disposition of that Kingdome or Country, into which it hath bene already, or is to be brought."[54]

The awareness of change and alteration and the consciousness of time intensified in the early seventeenth century. As men's goals became more secular and temporal, and as traditional means to felicity and redemption underwent redefinition, theorists became more concerned with understanding and directing time than with resisting it. Bacon and Raleigh expressed different appreciations of the relationship between time and order in the world.

Raleigh addressed the question of time's affect on the historical world much as Machiavelli had, but his writings lacked the fervent sense of hope and purpose that characterized the Florentine's. Man might act and shape things in the world but finally, Raleigh believed, man could not eschew destiny, and destiny meant decay. "There is nothing exempt from the Peril of Mutation; the Earth, Heavens, and the Whole World is thereunto subject," he bemoaned.[55] His fear of time and decay led him to resuscitate the ancient concept of cycles, as had Italian theorists such as Machiavelli. In so doing, he avoided placing the full burden of responsibility for social and historical change upon man and preserved a providential sense of first principles.[56] Man was only a secondary cause of civil war, he explained. "The first *Cause* of Civil War proceedeth of Destiny."[57]

Despite this concept of cycles and "Destiny," Raleigh's writings about particular real-political issues evidenced a conspicuous disregard for first causes. Consequently, his reliance upon such concepts appears to be rhetorically motivated. Raleigh appreciated man's potential, but he was sensitive to the emotional encumbrances that accompanied the consciousness of self-responsibility. By finally subjecting the necessities and the actions of reality, of man in history, to a historical first cause and Providence, he facilitated man's emotional and ideological emancipation from traditional concepts and smoothed the way to a redefinition of man's relationship to God and society. Raleigh recognized that mutability was natural, and he counseled flexibility; but he possessed neither Hobbes' idea of motive political psychology nor Shakespeare's vision of complementarity. Thus his rather tense and haunting sense of impending decay: "All Commonwealths alter from Order to Disorder, from Disorder to Order again; for Nature having made all worldly things variable, so soon as they have attained their utmost Perfection and Height, they must descend."[58]

Bacon, on the other hand, challenged time and fortune. While Raleigh was wary of innovation and new laws, Bacon urged princes and "Estates" to act to improve society, "for by introducing . . . ordinances, constitutions, and customs . . . they may sow greatness to their posterity and succession. But these things are commonly not observed, but left to take their chance."[59] Bacon contemplated time and fortune, too, but he refused to retreat under pressure from them. Man had to act to confront time.

Time generally corrupted things, Bacon thought. And though he argued against change for its own sake, he noted that received ideas and customs were insufficient guides to action. For Bacon, man had to alter things in order to stave off the corruption of time. Use change, he exhorted, to resist change, "for it is true that with all wise and moderate persons custom and usage obtaineth that reverence, as it is sufficient matter to move them to make a standard to discover and take a view; but it is no warrent to guide or conduct them; a just ground I say it is of deliberation, but not of direction."[60] Bacon appreciated policy and human action as he appreciated knowledge—they were practical and necessary means to social utility, and they armed man in his battle to overcome the ravages of time nontragically:

> Who knoweth not that time is truly compared to a stream, that carrieth down fresh and pure waters into that salt sea of corruption which environeth all human actions? And therefore if man shall not by his industry, virtue, and policy, as it were the oar row against the stream and inclination of time, all institutions and ordinances, be they never so pure, will corrupt and degenerate.[61]

Clearly man must retard and resist decay. Bacon's metaphor expressed confrontation; man must act, but he is not acting in an open-ended world yet. In reality his man reacts positively within the already closed, full cosmos; but by self-conscious action and direction, he redefines the structure of the world and usurps responsibility for its shape. And to facilitate this confrontation, Bacon exorcised the haunting fear which gripped his time that any change, "though it be in taking away abuses," will eventually "undermine the stability even of that which is sound and good."[62]

"Innovations . . . are the births of time."[63] Bacon began to suggest that only with controlled change could man insure stability and order in society and in history. And he hinted at an awareness that in reality man, not time, fathered decay. Time marked the changes that men either affected or did not resist. He understood that received ideas decayed because men refused to accept them. And to blame time for the erosion of order was suspect. Indeed, he acknowledged the inherently radical tenor of even the most conservative rhetoricians, those who exhorted others to resist all change. Bacon argued that since order and history were marked by flux "a froward retention of custom is as turbulent a thing as an innovation."[64]

Within the tradition of the Renaissance neoplatonic cosmology, Bacon saw that "matter is in a perpetual flux, and never at a stay."[65] The world contained fixed totality that was ever undergoing redefinition. Change was endless, but the changed was full and finite; and man could manage to direct the endless change. "Time is the measure of business," he claimed.[66] Time, then, was purposeful; and action toward a goal characterized time as well as flux did. Man must act within time; he must change, shape, and order the world.

Bacon complemented his neoplatonic sensibilities with an appreciation of a full Democritean materialistic cosmos. And he derived his idea of the state, and ultimately of order, too, from this understanding. In *De Sapientia Veterum*, he emphasized the study of fables and pre-Socratic wisdom. Fables represented a middle ground between "hidden depths of antiquity and the days of tradition and evidence that followed," he maintained; they separated "what has perished from what has survived."[67] Among the finite elements of the full cosmos were wisdom and knowledge, and Bacon argued that "beneath no small number of the fables of the ancient poets there lay from the very beginning a mystery and an allegory."[68]

Such an identification of ancient wisdom with cosmic meaning owed much to the neoplatonic-Hermetic tradition associated with Pico. Yet by placing knowledge and wisdom within the finite, revealed, and mysterious cosmos, Bacon attempted to recover order without recourse to isolating self-consciousness. Unlike Shakespeare, Bacon resisted the temptation of transcendence. The fables revealed, through allegory, the way of secular life. They answered questions about state, counsel, rebellion, sedition, and social order.[69] Bacon, then, strived to define the secular, civil world that he understood to be the center of man's life in the same way that traditional theorists had defined and mythologized the divine cosmos around which their lives revolved. In this way, he smoothed the way to an acceptance of and a self-conscious responsibility for this secular world by identifying it with traditional, comfortable analogs. He helped to rationalize a consciousness of self that Shakespeare made historical through drama and that Hobbes tied to the creative process of political definition.

The continuing emphasis placed upon practical behavior and secular values during the early seventeenth century undermined the unself-consciously coercive nature of social order, while it contributed to the redefinition of the idea of identity and the meaning of order. Bacon read the fable *Styx* as an allegory about treaties and oaths. Such contracts which secured social order were adopted and kept for only one reason, he observed. This faith is "not any celestial divinity. This is Necessity (the great god of the powerful), and peril of state, and communion of interest."[70]

Practical business truths were separate from theological or philosophical truths, Bacon declared.[71] One had to judge everyday political behavior, for example, against its own set of practical utilitarian rules, and not against any prescribed, all-encompassing ethic. He agreed with other theorists that political and social ills derived from political and social problems. Poverty and consequent discontentment led to sedition, he said. "And let no prince measure the danger of them by this, whether they be just or unjust."[72] Henry Wright also studied the reasons for sedition. Oppression, poverty, scarcity, and fear contributed greatly to social disorder, he concluded.[73] Early seventeenth-century writers deemphasized Tudor commonplace arguments about order, degree, pride, and ambition when they counseled about the security of the civil society.

Both Raleigh and William Fulbeck agreed that money was the primary commodity necessary for maintaining stability and strength in the commonwealth. Fulbeck maintained that "money is the strength and sinew of a state."[74]

Bacon realized that no general remedies for social disorder existed. Each particular disease must be treated according to its own peculiar symptoms. There were, however, certain well-advised preventative policies that any sound government should pursue. Bacon appreciated the benefit of sound economic health and wealth and he knew that this required more than the accumulation of land or money. He favored an open and well-balanced trade and the "cherishing of manufactures." He called for price regulation, moderation in taxes, and "improvement and husbanding of soil." Any good policy, he believed, "is to be used that the treasure and monies in a state be not gathered into few hands."[75]

Throughout his great *Essays*, Bacon discussed the practical and realistic concerns of princes and commoners alike. He argued elsewhere that princes "are best interpreted by their natures, and private persons by their ends."[76] Like Hobbes, he advocated a careful, nonbrooding appraisal of one's nature, habits, and likely actions as the best means of understanding politics and business.[77] Introspection and the understanding of human nature led to success in temporal endeavors, Bacon claimed, and his essays guided readers in this necessary investigation.[78] He went beyond Montaigne and attempted to help man use his self-knowledge in the social, political world.

As did Shakespeare's characters Brutus and Hamlet, Bacon realized the essence of good timing. For profitable action one needed to appraise the "ripeness or unripeness" of a situation; yet Bacon always advocated the readiness to act: "For when things are once come to the execution, there is no secrecy comparable to celerity."[79] Flexibility was as equally important as was timing. If a man courted success in the world, he must be willing to change his opinions and his actions because "nothing is more politic than to make the wheels of our mind concentric and voluble with the wheels of fortune."[80] Bacon further advised his readers about the respective values of spontaneity and planning; he discussed the benefits of honesty and the benefits of dissimulation, too. But each indvidual's personal nature ultimately determined which strategy or course of action would lead him to success. For Bacon, "the well understanding and discerning of a man's self" was prerequisite to acting in and shaping the civil world.[81]

Others saw the world and man from a similar vantage. In his *Essaies Politicke and Morall*, Tuvil noted the importance of teaching and persuading by deed and action as well as by ideal and words.[82] He warned that man could not judge things according to preconceived ideas of justice or expect things always to be reasonable. Man must be "politic" and like Ulysses "rest undoutedly assured, that where *Reason* cannot prevaile, *affection* wil."[83] Tuvil's psychological understanding of man's behavior supported a sense of policy which resembled that advocated and used by Edmund in *King Lear*. Tuvil taught that "he that can handle men aright in their affections & knowes at what times, in what manner, and by what meanes they may best be stirred up, may rest assured, that before his mind be thoroughly known, he is alreadie Mauster of what his heart desireth."[84] And in agreement with Bacon, he argued that an understanding of man and an appreciation of the particular nature of things facilitated the successful consummation of individual endeavors. He advised his readers "to be guided a little by *Observation*,"[85] but he recognized that there "is nothing more hard, and difficult to come by then a true and certaine knowledge of the inward disposition, and abilities of man."[86]

Tuvil based his account of policy and action upon an embryonic, voluntaristic psychology of human ethics which indicated Hobbesian insights in many places. His understanding of the relationship between "passions," "will," and "reason" undercut the commonplace valuation of behavior that emanated from the concept of "right reason" and the Tudor idea of order, and required a reappraisal of the purpose and ultimate ends of private and public activity.

> Passions are certaine internal acts, and operations of our soul, which being joyned and linked in a most inviolable, and long continued friendship with the sensitive power, and facultie thereof, doe conspire together like disobedient and rebelious subjects, to shake off the yoake of *Reason*, and exempt themselves from her commaund & controlement, that they may still exercise those disordered motions, in this contract world of our frayle and humaine bodies, which during her weaker Nonage, they were accustomed to doe.[87]

This conspiracy employed "wit" and "will" to suppress "reason" and bribed her with pleasure, Tuvil continued.[88] Man's desires reflected the control of his passions:

> The *Wit* . . . labours to find out reasons presently, that may countenance & grace it [the desired object], and the Imagination . . . like a deceitful Counsellor . . . represents them to the Understanding in a most intensive manner.[89]

The internal motivations that directed man's actions paralleled the motivations that directed public policy. The knowledge of man led to the knowledge of society. No longer were the body politic and the human body correspondent to natural order and divinity. In Tuvil's world of motive behavior and practical policy, man and the state were analogs for each other; and man was becoming the primary referent. In a fashion that looked ahead to Hobbes' concepts of "appetites" and "aversions," Tuvil recognized that behavior was not motivated by externally defined ethical purposes, but rather by simple internal passions: "tis eyther hope of *Reward*, or feare of Punishment, that in the attempt of things, orders and directs our choyce."[90]

Bacon was not blind to the need to examine the ethical premises for meaning and behavior. Value measured the capacity of something to facilitate desired purposes, he contended. As Shakespeare did in *Troilus and Cressida*, Bacon sought to fix the locus of meaning; was anything meaningful in itself? Even as a young man, he realized that received ideas concerning meaning required modification. In *Of The Colours of Good & Evil*, written in 1597, he admitted that a proper knowledge of value cannot be discovered "but out of a very universal knowledge of the nature of things."[91] Meaning was particularly defined, then, and man's behavior and his purposes and goals in life were particular, too. Bacon advised man to "judge of the proportion or value of things as they conduce and are material to . . . particular ends."[92]

Certainly such an understanding of the particular and the purposive nature of meaning, and such an acceptance of man's ultimate responsibility for his behavior, affected man's concept of self and his understanding of his relationship to society. Identity could no longer easily be described as a correspondent by-product of the divinely ordered cosmos. Now man began to seek meaning and identity in the secular world, and in doing so, began more self-consciously to experience emotional isolation and a sense of cultural decay. From a psychological as well as a social perspective, then, the process of meaning redefinition, almost of necessity, was characterized by an attempt to establish similar self-consciously defined ends for man and for society.

Yet the conciliation of self and society, of private and public meaning was a difficult task; indeed, ambivalence characterized the effort. Early seventeenth-century Englishmen glorified the individual but they feared him, too, and they attempted to resist the uncertainty that surrounded him. And the feeling that private endeavors were inherently antipublic remained. Raleigh instructed his son and posterity that man is fickle. Love God, your country, your prince, and your own estate "before all others, for the fancies of men change and he that loves today hateth tomorrow," he warned.[93] Raleigh feared flux and uncertainty and sought security in institutions rather than in man; but he also sensed that man and society were separate and that man had to identify himself socially rather than privately.

Bacon struggled with the private-public relationship, too. Perhaps no other contemporary writer, besides Shakespeare, recongized and praised man's responsibility for social order as did Bacon. Time, fortune, and natural tendencies did not contrive the shape of things; individuals did, he argued. Speaking about the contemporary political order, he declared, "this peace so cultivated and maintained by Elizabeth is matter of admiration; namely that it proceeded not from any inclination of the times to peace, but from her own prudence and good management."[94] And in more general terms, about man's responsibility for fortune's whims, he maintained that "it is not moneys that are the sinews of fortune, but it is the sinews and steel of men's minds, wit, courage, audacity, resolution, temper, industry, and the like."[95]

But Bacon was wary of private aggrandizement, and he often doubted whether a lasting private-public relationship, which defined mutual goals, could be accomplished. In "Of Great Place" the public world and the private world appeared as antagonistic forces. Public men were obliged to forget their private selves. "It is a strange desire, to seek power and to lose liberty; or to seek power over others and to lose power over a man's self," he said.[96] Yet public purpose authenticated private aspiration. Power must be sought because it legitimized a man's actions; it was a license to do good or evil. And "power to do good is the true and lawful end of aspiring," he argued.[97] Unlike Hobbes, Bacon could not praise aspiration for its own sake. Hobbes believed that good and bad actions were similarly motivated; Bacon sought to separate them. Public good, not self-aggrandizement, should inspire private desires. Bacon sought public definition for private action even while he sensed that they were inimical, because, though he celebrated man's responsibility for his social world, he feared, too, the tragically isolating consequences of self-definition.

In "Of Wisdom For A Man's Self," Bacon modified the typical humanist advice about the relationship of self and society. Bacon sensed what Shakespeare's Polonius did not realize: that total self-definition isolated man. "Divide with reason between self-love and society: and be so true to thyself, as thou be not false to others."[98] While he appreciated the possibilities of self-responsibility,

Bacon also shared Shakespeare's vision of the tragic consequences that accompanied a radical reliance upon self; and he hoped to escape them. Cordelia and Coriolanus finally sacrificed themselves to their own self-truths. And Bacon knew that those who relied totally upon themselves, or who loved themselves above all others, "become in the end themselves sacrifices to the inconstancy of fortune; whose wings they thought by their self-wisdom to have pinioned."[99] This anxiety about the relationship between self and society characterized much of Bacon's work and gave it a dramatic quality that intimated the same world as did Shakespeare's plays. Like Shakespeare, Bacon understood the potentially tragic and negating consequences of self-definition. Consequently, by advancing utilitarian ends for private individual actions, he attempted to conceptualize identity in a fashion inclusive enough to conciliate private and public meaning.

Between 1628 and 1632 the incarcerated Sir John Eliot wrote two pamphlets which, in the development of their arguments and themes, reiterated many of the changing conceptions about man, society, and order that characterized the tenor of intellectual and political thought in the first half of the seventeenth century. Eliot's work borrowed commonplace attitudes from traditional Tudor order theory while simultaneously discussing and defining concepts of authority and sovereignty much in the manner of Bacon and, oftentimes too, of Hobbes. The earlier of the two pamphlets, *De Jure Majestatis*, was primarily a political tract that labored to establish a relationship between authority, law, sovereignty, and order.[100] Received concepts, identifiable with the Tudor idea of order blended, here, with more particular developing concepts of secular sovereignty. In this treatise, Eliot focused attention upon the "Public Good" and indicated the explanatory power of natural analogy, and as did Bacon and Raleigh, he deemphasized biblical and religious argument and explanation.

During the early decades of the seventeenth-century theorists and politicians addressed the subject of sovereignty: its definition and its limitations. Eliot began *De Jure* by arguing that sovereignty implied supreme power.[101] He allowed that individual kings possessed this sovereignty, and that nature affirmed this: "Nature doth indeed teach us subjection & pointe at Monarchie, but in every kingdome."[102] But he also maintained that absolute power and majesty existed independently of the title of kingship.[103]

Eliot continued his inquiry into sovereignty and authority by viewing them in relation to society and order. He understood order in terms of traditional concepts and commonplaces marked by correspondence theory, but he discussed society in terms of the "State," the end purpose of which was to order the society. As Eliot said, "A commonwealth is an order of a State. Order must begin at one first or chiefe which must be first either in nature or in Analogie. . . . The chiefe is that which in that order hath no superiors."[104] Since the chief aim of order is government, he continued, that which rules all and is governed by none is chief.[105] Order could be obtained, he argued, only if power or sovereignty remained undivided. Limited supremacy "destroys the nature of order which alwais requires one chiefe, for the preventing of confusion which must necessarily follow such divided government."[106] These views suggested a theory of sovereignty that Hobbes more fully elaborated a few years later. But Eliot displayed none of

Hobbes' political psychology here. He discursively organized his arguments in characteristically commonplace fashion. And, except that he explained sovereignty without mentioning divinity, to this point, Eliot certainly said nothing that would have angered Elizabeth, James I, or even Charles I.

Eliot began, however, to limit absolute authority by discussing definitively the theory of sovereignty in relation to the king, law, and order. People do not make kings, he affirmed; and then he qualified this view by arguing that even if they do, as some claimed, "once they have conferred their right upon another they have deprived themselves of it."[107] Nonetheless, the king should govern by law and reason, not by will and lust, he maintained. If he does not, however, "he is not to be judged by humane laws."[108]

At this point Eliot began to weave a subtle web in which he rhetorically related the king's public and private persons. The king should willingly limit his own power. While he is not bound to live according to civil laws, said Eliot, he should place himself under the law, "for then is he *Rex Regis*, a King over himselfe."[109] Following this logic, Eliot cunningly differentiated between a tyrant and a king: "The one usurpes more authority than he should, the other doth not exercise all the power that he might."[110] A true king, then, just as his subjects did, voluntarily limited his own possibilities. Potential, rather than kinetic, energy characterized Eliot's concept of sovereignty. Consequently, he argued that a king "should not use his authority when he may use lawe."[111] And though Eliot had argued that law derived from authority, he now separated the two. Once law existed, it acquired a sovereignty, or at least a limiting authority, entirely its own.

Near the end of *De Jure*, Eliot introduced the issue of the public good into his discussion. He refuted the premise that the king owned all the property in the realm and defended the notion of private property. But he argued that in times of extreme necessity, when the public good demanded it, the king could "lay hands upon private goods."[112] But Eliot neither defined what the public good entailed, nor who was to decide what was or was not in the interest of the public good. Yet this established a precedent because the public good became the point of final arbitration of many issues; and different sources of authority claimed to interpret and to represent this ubiquitously espoused yet ambiguously defined concept. Eliot concluded that "a Prince therefore may not alienate either the whole Common wealth, or any pat of it, without the consent of those, who have interest in the propriety, as he hath in the sovereignty."[113] Evidently then, the sovereign had voluntarily to limit his prerogative and his authority and had to balance them against certain natural rights of propriety which were becoming evermore self-consciously represented.

The Monarchy Of Man was a far more reflective work than was *De Jure*.[114] Both a philosophical disquisition and a political treatise, *The Monarchy Of Man* seriously considered the nature of man in a changing political world. Eliot still discoursed about man, the microcosm, by correspondence analogy, but now he likened man to the state rather than to nature, intending to describe man's particular "frame and constitution." Indeed, man himself reflected God's greatness: "Man to be ye governor of himselfe, an exact Monarchie within him, in the composition of which state, nothing without him may have interest, but all stand as subservient to his use, hee only to his Maker."[115] And the consequent purpose of both, man and state, or as Eliot had it, subject and sovereign, was a shared common duty to the "public utility and good."[116]

For Eliot, divinity ordered experience uniformly. All cause and effect emanated from the one order. He did not appreciate with Bacon or with Hobbes the defining nature of motion or of appetite. But he acknowledged a changed political framework in which this order resided. *The Monarchy* attempted to stretch the metaphorical assumptions of *De Jure* to their practical conclusions. This was especially the case regarding the interrelatedness of law, authority, and monarchy, and regarding, too, the consequent shared responsibility of subject and monarch for the public good.

Borrowing from Plato and, interestingly, from Bodin, Eliot denied that kings were absolute and could sin only against God. In the tradition of Hooker, too, he concluded that "nothing but ruin can be the fortune of that Kingdome where the Prince does rule the lawes, & not the lawes the Prince."[117] He had shifted position. From what authority did law spring? Now law seemed supreme. It was no longer a question of whether laws influence kings; now one had to measure how influential law was, and, tangentially, one had to know wherein law was represented. Conclusively then, Eliot affirmed that "it is necessary by the wisdome of a Counsell, Lawes should be prescribed, that by the fourme & rule, all Judgements might be regulated."[118]

Eliot's concern about the responsibility of both king and subject for the public weal fashioned the core of this revised appraisal of the relationship between law and authority. "Law is the ground of all authority," he now maintained, and both the subject and the sovereign bore an equal duty to and obligation for this authority,[119] because they shared a common bond of safety.[120] And since power was rooted in will, men as kings must be "inferior to themselves"; the good king would be less than he could be.[121]

In thus limiting the monarch's authority Eliot continued the rhetorical argument he began near the end of *De Jure*. But, more interestingly, he began to articulate legislative sovereignty as discourse by conceptually redefining the meaning of social responsibility. Of all thing considered for the commonweal, Eliot argued, no matter the source, "Princes onely can reduce them into act; & by the influence of their power give them . . . both spirit & perfection."[122] Eliot did not refer to the exercise of power, but to the influence of power. The king's will reduced or defined the subject's will metonymically. It is in this sense that "framing the constitution" must be understood. The king merely articulated the projected desires of the subjects who shared with him the responsibility for public utility. And since no king could be omniscient, the subjects must advise the king. To know, the king required counsel, and understandably then, the king's council, or Parliament, authorized, and elevated the acts of princes.[123] Consequently, "the health and safety of the Senate" strengthened the subsistence of the state and the public weal.[124]

A self-consciously conservative posture characterized Eliot's philosophical and psychological considerations as well as it did his political conclusions. As did so many of his contemporaries, Eliot burdened his political treatise with weighty discussions about felicity, motion, and cycles. Although the end of good government was order and public utility, still, he claimed, government was merely the medium to facilitate man's self-contentment, his "perfect felicitie."[125] And, he argued, reclamation, not innovation, was the best means to this end. Man contained, within himself, the natural principles necessary to secure *summum bonum*, and his effort must focus upon ridding himself of the corruptions that inhibited his natural potentialities.[126]

Man could best oppose self-corruption by working through the state.[127] Contrariety characterized man, society, and nature. As soon as man understood this, Eliot maintained, he could manage his contentment. Indeed, sorrow, the worst corruptor of man's natural propensities resulted from man's inability to accept that the world was in "continuall revolution." Man should locate peace in the state rather than continually confronting uncertainty and mutability.[128]

It was by way of this understanding of "continuall revolution" that Eliot articulated his developing appreciation of man's responsibility for social change. The characteristic conservative revolutionary, he defended innovation as renewal and explained all change by recourse to correspondence theory and to providential cycles. As did Raleigh, he understood that social unrest reflected historical cycles, but he believed that man could shape social change. "Warre is the cause of peace," he argued.[129] Thus he justified the utilitarian benefits of social unrest and, at the same time, he anchored social change to the firm foundation of providential cause.[130] Metaphorically he related the physics of social and political change. Just as was the planetary order, the social world, too, was characterized by a variety of motions fulfilling one plan, he said.[131]

Eliot used traditional rhetorical discourse in a way, however, which transcended its limiting principles. In attempting to ease the perturbations of man's mind which Eliot appreciated as resulting from man's confronting a changing society, he imaginatively modified the metaphorical foundation for the idea of order. For Eliot, man could take responsibility for initiating what was natural. Man's quest, he maintained, must be to find the way to instigate the renewal.[132] Synechdochically, he integrated nature's cycle with particular human innovation. Renewal was naturally defined and socially articulated. Man acted to experience that which was natural. Consequently, change reflected the providential plan, but it was a means to facilitate specific, advantageous ends as well.

Action, then, was both publicly prescribed and privately oriented. Man could bend his will to reason and could direct himself to Ithaca, to good government, Eliot concluded.[133] He placed the public good above any private good and expected that man actively worked for the utility and profit of the many. Indeed, as for Bacon, for Eliot the edification of the self prepared man to benefit others.[134]

But since action was also the "End of Speculation," the end of man's activity was similarly "Contemplation."[135] As Bacon did in *New Atlantis*, Eliot turned finally to the natural philosophy associated with Marsilio Ficino, Pico, and John Dee to help him understand man's place in the changing world. Only the activity of man's mind captured the essential qualities of innovation and renewal which characterized cosmic motion, Eliot believed. "Contemplation in all actions makes as well the end as the beginning," he noted, referring to Ficino.[136] In the tradition of Dee, of Prospero, and of Bacon, Eliot celebrated and deified the mind's active powers of motion and of operation. The mind, he claimed, "breaks through the orbes and immense circles of the Heavens, and penetrates even to the center of the Earth."[137] He no longer sought correspondent meaning for man's activity in the state or in the monarch, but he understood, rather, that meaning resided within man himself. All things, places, and elements were "in one thought . . . within the comprehension of the mind."[138]

The contemplative mind of man actively influenced existence. It knew no restrictions and "it has libertie upon all."[139] Consquently, Eliot concluded that man

> is an absolute Master of himselfe, his owne safety & tranquility by God . . . are made dependent of himselfe. . . . And in that selfe dependence . . . is that *summum bonum* . . . the true end and object of this Monarchy of Man.[140]

This was surely strong stuff from the conservatively oriented Eliot. Understanding, as he did, that man's responsibility for order in a changing world required oriented action, Eliot, as did Bacon, borrowed theories about man's operational powers from a growing philosophical tradition. In this way Eliot's man became master of his own fate, but he did so within the traditionally defined cosmic order.

Before considering the nature and the extent of this philosophical tradition, it would be useful to examine mercantile values, merchant pamphlets, and attitudes about merchants in the late sixteenth and early seventeenth centuries. Joyce Appleby suggests that we should approach the intellectual response to economic crises and change during this period "as a creative social act. . . . The initial effort to explain the wide-ranging influence of the free market became a part of the secular ideology that replaced traditional social assumptions in England by the end of the [seventeenth] century."[141] And in a controversial study, W. K. Jordan has argued that the merchant elite of this period helped fashion the new ethic of social responsibility.

Merchants' aspirations were more farsighted, more secular, and more concerned with social utility than were those of any other distinct social group, Jordan claimed. And since the "men of the sixteenth century had of necessity to begin to build their institutions *de novo*,"[142] merchants indicate the value shift from primarily religious preoccupations to secular concerns. And by differentiating "things economic from their social contexts," the mercantile writings of this period contributed historically to ideological redefinition.[143]

Economic crisis and social instability during the middle years of the sixteenth century and again during the first decades of the seventeenth century prompted a spate of pamphlet literature which discussed the economy in general and merchants in particular. A careful reading of many of the merchant pamphlets suggests that they were not merely responses to economic crises; they were also expressions of growing merchant class self-consciousness. While this literature analyzed the contemporary economic malaise, it also iterated respect for the merchant, his vocation, and the worth of commerce, and defended secular values regarding economic as well as other aspects of life. Again, to cite Appleby's fine study: "To respond positively to the opportunities for further economic development . . . required the endorsement of new values, the acknowledgement of new occupations and the reassessment of the obligations of the individual to society."[144]

It was during these years that men began to be quantitatively minded.[145] They became more conscious of numbers as well as of time. As early as 1549, Thomas Smith pictured the economy as a mechanistic system, removed from the divine order.[146] Personal and temporal goals and explanations influenced economic discussions, too. Even Gerard de Malynes, a conservative theorist who still described the economy in geometric and mystical terms in the 1620s, argued that "gaine" was the center of the "circle of commerce."[147] Such value changes, John Nef maintained, "are of decisive importance in orienting economic endeavor in new directions."[148]

Secular values and attitudes received a sustained airing in mercantile literature during these years as writers constantly tried to justify commerce and the merchant to an unreceptive public. Well into the seventeenth century there still prevailed a "distrust of the self-made men. . . . The suspicion of business wealth as a social solvent and a certain lingering nostalgia for an ordered, traditional society was to die hard."[149] Consequently, in economic activity, as in politics, ambivalence was characteristic. And R. H. Tawney noted that "in practice since new . . . interests and novel ideas had arisen but had not yet submerged their predecessors, every shade of attitude . . . was represented in the economic ethos of the age."[150]

No matter how affluent or powerful the merchant was, he rarely acquired equivalent social status. The medieval merchant was individually isolated and socially insecure. He joined with other merchants in "societies," seeking security through fraternity.[151] Within these societies a system of ceremony and respect based upon hierarchy and order and resembling the larger social order prevailed.[152] Throughout the Middle Ages "no alternative system of valuation gained currency. The cult of secular equality . . . found no prophets."[153] Generally, merchant insecurity and social alienation continued throughout the medieval period. A large majority of London's merchants, for example, associated only with other merchants, and entrusted only to other merchants the care of their children, the safekeeping of their valuables, and the administration of their lands.[154] Whenever possible, however, merchants or their heirs moved onto the land and assumed the gentry life-style. As a result, merchants gained the reputation for being socially ambitious.[155] Yet it was disdain for the merchant in the first place that prompted so many merchants to seek landed status. "Chivalric literature . . . singled out the merchant as the type of avarice and ambition . . . in contrast to the noble who . . . made life an art."[156]

Seventeenth-century merchants railed against the prevalence of these attitudes. They were still anathematized by many as "a marvelous destruction to the whole realm."[157] Sir Thomas Smith's *Discourse* of 1549 was in many ways a harbinger of much of the economic literature written sixty and seventy years later. The *Discourse* discussed the mid-century economic crisis in a surprisingly sophisticated fashion.[158] Implicit in the dialogue, nevertheless, were many of the then-current attitudes about the merchant. Smith suggested that in any economic crisis, merchants would gain most, while noblemen and gentlemen would lose most. Merchant men are dangerous to the commonwealth, he noted, and "can best save themselves in every alteration."[159]

Twenty years later, Thomas Wilson wrote his famous dialogue *A Discourse Upon Usury* which Queen Elizabeth supposedly sanctioned. Wilson managed to include in this treatise many of the contemporary conflicting values and attitudes about commerce and the merchant. Although the "merchant" enunciated his position quite forcefully in the dialogue, finally he was convinced that his views, and by suggestion, those of the merchant class, were misguided, and he recanted. Throughout he was addressed as "Gromelgayner." "They term you by the name Gromell gayner, because you mynde nothyng so much as gettinge of money."[160] And herein resided one of the main reasons for social discord: "For men of wealth are now wholly geeven everywhere all together to idleness, to gett their gaine with ease, and to lyve by lending upon the onely sweate and labour of others."[161]

Even though merchant wealth benefited the monarchy, James I had few good words for merchants, who "regard the commonwealth as existing to serve their turn, and do not scruple to feather their nests to the public loss."[162] Indeed, in his advice to his son, James tried to foist entire culpability for economic crises upon merchants. "The Merchants thinke the whole commonweale ordeined for making them up," he complained. They aimed to enrich themselves at the expense of all the rest. They export necessary goods and import superfluous goods, if they import any at all, he continued. "They buy for us the worst wares, and sell them at the dearest prices," and "they are also the speciall cause of the corruption of the coyne, transporting all our owne, and bringing in forraine, upon what price they please to set on it."[163] Later in the century, Thomas Hobbes, surely a man comfortable with secular values, maintained that "the only glory" of merchants was "to grow excessively rich by the wisdom of buying and selling."[164] Such sentiments were more the rule than the exception, and while the merchant experienced greater social mobility in the seventeenth century than he had in the Middle Ages, he often still sought social acceptance and status by emulating landed class behavior. Lionel Cranfield, one of the most prominent merchants in Stuart England, who, in fact, became Lord Treasurer, exemplified the aspirations of his class. Cranfield aspired to become a "merchant prince."[165] His great house in London, intended as a "palace," epitomized the conspicuous consumption common among London merchants.

Socially stigmatized, the merchants of early seventeenth-century England attempted to validate a growing self-confidence by joining new personal values with public policy. As Professor Jordan found, "merchants . . . were still . . . uncertain of their status. Yet the very fact of their great wealth . . . and their willingness to assume an immense burden of social responsibility created for them the reality of status."[166] Conscious of the fact that the traditional foundations of society were weakening and confronted with sustained commercial crisis, the "merchants became aware that they could actually influence policy, and they begin to search for rationalizations that would give cogency to their demands."[167]

Perhaps the best example of the self-conscious and self-confident merchant was Edward Misselden. In his pamphlets, *Free Trade*, published in 1622, and *The Circle of Commerce*, published in 1623, he not only advocated the worth and the dignity of commerce and of the merchant, but he posited a conciliation between the merchant's private values and life and the public weal. Misselden dedicated *The Circle of Commerce* to Cranfield, "the Mirror of Merchants, the Luster of

London."[168] After politely honoring the king and the nobility, and asking Cranfield to "make much of Merchants," Misselden assumed responsibility as a merchant to suggest ways to save trade, and, thereby, the kingdom.[169] "I spoke with my pen, as I never spoke before . . . O men kill not the Kingdom!"[170] In *Free Trade*, dedicated to Prince Charles, he affirmed that "surely matters of State and of Trade are involved and wrapt up together."[171] He regarded it as his duty to God and to kingdom, he declared, "to express the same in some publique service for the publique good."[172] The merchant owed it to his country to solve the commercial crisis, Misselden maintained. Such was his incumbent responsibility, for "what hath more relation to matter of State, then Commerce or Merchants. For when Trade flourisheth the King's revenue is augmented, Lands and Rents improved, Navigation is encreased, the poor employed. But if Trade decay, all these decline with it."[173]

Misselden refuted the charges of Malynes and others that merchants would not make good government advisors. Malynes differentiated between the merchant's private interests and the public good, and he also suggested that, often, merchants were ignorant about commerical matters.[174] In response, Misselden asked:

> Is it not lawful for Merchants to seeke their *Privatum Comodium* in the exercise of their calling? Is not gaine the end of trade? Is not the publique involved in the private and the private in the publique? What else makes a Common-wealth, but the private wealth, if I may so say, of the members thereof in the exercise of Commerce amongst themselves, and with forraine Nations?

And he concluded, then, that "the wealth of every Commonwealth, hath a Correlation with this Nobel Profession."[175] He made another case for the merchant's value to society, and he expressed his self-confidence and his awareness of the merchant's capacity to deal with a changing world when he confronted Malynes' charge about merchants' ignorance. "The merchant is the best to consider commoditie quality and balancing of trade," he avowed; "Who is more fit then . . . merchants to unmask the mysteries of *Maunte Bankes*, Iugglers, and Imposters of Trade?" And in an even broader reference he concluded: "Merchants I say, besides their knowledge of Commodities, and . . . Exchanges . . . are acquainted with the Manners, Customs, Language, and Laws of forraine Nations. . . . theres none more fit to make a minister for a King, then an expert and iudicious Merchant."[176]

The desire and attempt to earn respect for the merchant, and the justifications necessary to sanction the merchant's public worth, helped to legitimize socially the secular values identified with commerce and with merchants, and helped to redefine the meaning of economic activity. In order to justify claims of merchant value, Misselden, and others, constantly confronted current commercial problems and sought new means to economic and social relief. Still, Misselden's ideas were characteristically ambivalent. While he defended private, secular desires and gain as suitable purposes in life, he understood commerce as an aspect of a larger cosmic order. As did Malynes, he spoke in geometric terms, and he defined trade and exchange with Aristotelian conceptions of action, passion, form, matter, and deprivation.[177] Though he described commerce traditionally, he transcended the old order in regard to the place of the merchant in society. In this way he helped to free the merchant from the rigid structure of the traditional social hierarchies, and helped to create a new meaning of economic activity.

Misselden was concerned about merchant honor and security, too. Although confident, himself, of the merchant's worth and abilities, he realized that, because of social disapprobation, many merchants' sons chose not to follow their fathers' professions. Misselden argued that they should, because

> merchants are wont to make it their glory, to advance their fortunes, renowne their names . . . and to perpetuate the same unto posteritie, as an hereditary title of honor unto their name and blood. . . . For where the father doth thus ingenerate his sonne, and the sonne doth not degenerate from his father, there the Estate is kept entire, in its owne stock: there the spring doth not spread it selfe into stragling streams in which their fame is lost, their name put out, and the Estate consum'd in ryot: and this is a Common losse unto our Commonwealth.[178]

Again, Misselden treated the questions of merchant social status and of merchant interests as functions of the public benefits accrued from a healthy mercantile class.

Of the articulate mercantile authors, Misselden appears to have been one of the most psychologically perceptive. His lucid style and mode of argument encapsulated what was more disjointed in other writers' works, and his writings better withstand a thematic analysis. Nevertheless, all the issues with which Misselden was concerned—the merchant's desire for respect, the value of commerce and of the merchant for the commonwealth, the discussion and approbation of new values, the desire to generate filial loyalty in mercantile families, and, finally, the conciliation of private values and public welfare—were common themes in many other mercantile works and literatures treating merchants of this period.

In both Sir Thomas Smith's and Thomas Wilson's dialogues, merchants praised the notable acts of men of their profession in order to demonstrate their worth to the commonwealth. Merchants founded hospitals and gave relief to the poor. They "redeem the custom" of the city and supply the money for the local charges levied on the city.[179] Who else maintained the state by lending to needy princes? "Is not London the Queen's chamber? Are we not then chamberlens to her maiestie, people alwayes ready to spend not onely our goodes, but also our lyves in her service?"[180]

Despite what they saw as their conspicuous value to the commonwealth, merchants were well aware that they and their profession were continually denigrated. Consequently, during the first four decades of the seventeenth century, merchants advanced a variety of self-defenses and defenses of commerce. John Wheeler attempted to prove that trade and bartering were natural to the laws of man and society.[181] He recognized no conflict between the values of the traditional idea of order and society and the pursuit of commercial activity. Still he acknowledged that many merchants were ashamed of and were shamed by their profession. Commerce should not be despised, he declared, but "it is to be prooved that the estate is honorable, [and] may be exercised not only of those of the third estate . . . but also by the Nobles and Chiefes & men of this Realme with Commendable profite, and without anie derogation to their Nobilities, high degree and condition."[182] Commerce benefited the commonwealth, the prince, and the nobility, "without the loss of one jote of honor."[183] Wheeler was a traditionalist confronting change as well as he could. He defended merchant companies because they ordered commercial activity, and he attempted to reconcile the merchant profession with the established system of social hierarchy. Although he empathized with merchants who sought respect and higher social status, he refused to demand the same if it meant transcending the confines of the social structure.

Gerard de Malynes, another traditionalist merchant, also struggled with the question of merchant status. In his magnum opus, *The Antient Law-Merchant*, first printed in 1622, twenty years after Wheeler's treatise, he supported the assertion that "the state of a Merchant is of great dignitie and to bee cherished" by claiming ancient custom and heritage for the merchant.[184] As did so many politicians, political theorists, and legal theorists of the day, Malynes attempted to sanction what appeared new and upsetting to the social order by locating ancient precedent for it. Characteristically, then, he declared that the "*Law Merchant* hath always beene found, *semper eadem*, that is constant and permanent, without abrogation, according to . . . most auncient customes."[185]

Two other pamphleteers, Thomas Mun and Henry Robinson, rejected the ignominious reputation that merchants bore. Neither, however, sought status for the merchant in accord with the social commonplaces of the old order. Mun, in fact, expounded strictly practical reasons for honoring the "nobleness of this profession."[186] As long as society aspersed merchants and commerce, Mun argued, merchants would not be motivated to compete successfully with their Dutch counterparts:

> It is true indeed that many merchants here in England find less encouragement given to their profession than in other Countreys, and seeing themselves not so well esteemed as their *Noble Vocation* requireth . . . doe not therefore labour to attain unto the excellence of their profession.[187]

Robinson vindicated the social status of the commercial class by elaborating the functional similarities between aristocrat and merchant:

> Everie one, whether he will or no, is a merchant for what he buyes or sels, be it lands, houses or whatever else, and more gentle it is to sell cloth, silk, satins, as meere merchants doe, then cattell, hay, hides, wooll, butter, cheese, as countrey gentlemen, and others of best note and worth . . . have always done.[188]

Indeed, Robinson repudiated the traditional understanding of hierarchical social status. He proposed new means to evaluate worth and to ascribe status based upon utilitarian benefits to the commonwealth. Under such a system, one merchant should "be valued as hundreds of ordinarie men, because many hundreds of men are employed and maintained by one merchant."[189] Here Robinson expressed both an awareness of the merchant's important contribution to a society undergoing meaning redefinition, and a self-consciously positive sense of identity, bolstered by pride of occupation.

As did Misselden, many merchants in the early decades of the seventeenth century, whether traditionalists or not, asserted confidence in their own abilities to understand the economy and to devise plans with which they could counsel princes about the economy. Malynes subtitled his major work: "necessarie For All Statesmen, Iudges, Magistrates, Temporall and Civil Lawyers, Mintmen, Merchants, Mariners, and all others negotiating in all places of the world."[190] Mun confidently suggested ways to increase exports and limit imports. He advocated a conscious manipulation of manufacturing, and he displayed a wide-ranging knowledge of land cultivation, cost and price relationships, customs, and imposts.[191]

Lewes Roberts, however, devised an extensive plan for the facilitation of trade and merchant success. Roberts' scheme included all aspects of trade and of merchant relationships. He felt that society should "give honor to merchants," because commerce was a profession "where-in there is so many usefull and principall parts of a man required."[192] His demands included an agency to maintain the constancy and the quality of the coins, and a board of merchants authorized to establish all regulations concerning commercial enterprise.[193] Roberts' plan was more suited to a developing market economy, based upon success and riches, than to a traditional economy still ordered and defined by social hierarchy. He appreciated the importance of the individual, even within the context of the "common-weal," and the secular values which informed his plan insured identity, status, and respect for the merchant.

Roberts argued, as Misselden did, that merchants' sons should also become merchants. Mun, too, was adamant on this point. He was bitterly opposed to a son who would rather "consume [his father's] estate in dark ignorance," by becoming a gentleman, "than to follow the steps of his father as an Industrious Merchant to maintain and advance his Fortunes."[194] Roberts went even further and proposed that the sons of the nobility should enter trade. Both the individual and the commonwealth would benefit, he concluded. "And if Gentlemen in generall would thus apply themselves to traffike . . . they should by all likelihoods benefit themselves more in one yeare . . . then peradventure at Court by ten years waiting and solicitations."[195]

Merchants needed to argue their utility for society and most pamphleteers understood that the best way to do this was to relate commerce to social welfare. Wheeler asserted bluntly that "without Merchandise, no ease or commodious living continueth long in anie state, or commonwealth."[196] Even Wilson's dialogue expressed the view that the merchant's treasure and the welfare of the realm were closely related.[197] Lewes Roberts claimed that "traffike" was the safest way to "inrich a country,"[198] and that "a well ordered traffike managed by skillful Merchants, hath beene, and ever will be, honorable to that Kingdome and Soveraigne, where the same is duely practised, and carefully protected and preserved."[199] In one of his earliest pamphlets Malynes stressed the very practical benefits that society accrued from commerce. He specifically detailed that

> the more readie money either in specie or by exchange, that our merchants should make their returne by, the more employment would they make upon our home commodities, advancing the price thereof, which price would augment the quantity by setting more people on worke: and would also increase her magesties customes outward. All which is tending to the general good of her Magesty, the whole realme, and every inhabitant thereof.[200]

Even a pointed defense of the East India Company such as Digges published in 1615 inclined to such arguments. In a letter to Sir Thomas Smith, governor of the company, Digges claimed that commerce and trade "clothe and feede the poore, and give the willing man employment to gaine with them, and with the Commonwealth."[201]

Nonmerchants harped upon this theme, too. Fulbeck recognized that it "is good for everie Prince to have speciall care and regarde of maintaining merchandize, because by that means, not onelie things profitable are brought into a kingdome, but manie thinges are carried out to be sold; and exchanged for publick good."[202] And Bacon described merchants as *vena porta*, and he noted that "if they flourish not, a kingdom may have good limbs; but will have empty veins, and nourish little."[203]

By associating their calling with the public good, merchants also began to describe a positive relationship between the new secular values which influenced their trade and the public welfare. Thomas Mun expressed the wish that "the private gain may ever accompany the public good."[204] Roberts wrote to aid "the inlarging and benefitting of my country by traffike, and the advancement of the Merchants."[205] Henry Robinson, a more important social thinker than Misselden, Mun, or Roberts, discussed the relationship between private interest and the public welfare in a sophisticated fashion which evidenced the evolving appreciation of secular values. "Trade was to him the solvent that would loosen the ossified structure of the past that had stultified and constrained the forces which were enlarging the opportunities of mankind."[206] Robinson, writing in the 1640s, described a new social order in which a secularly conceived economy would "augment the wealth and power of the nation, and . . . enlarge the opportunities of the individual citizen."[207]

There can be no doubt that merchant tracts helped the newer values of the period receive expression. Smith's *Discourse* described the economy as "a mechanism of forces; impersonal and amoral in character, subject to analysis and manipulation by intelligent policy."[208] Smith understood that the economy functioned according to its own mechanistic laws rather than to the divine laws of cosmic order. In keeping with this view and with the growing awareness of time, Malynes described the economy: "We see how one thing driveth or enforceth another, like as in a clocke where there be many wheels."[209]

The wave of pamphlets responding to the mid-Tudor economic crisis introduced the argument that man was free to pursue private economic interests, and consequently to affect social welfare.[210] Here, too, Smith's *Discourse* prefigured later tracts in its recognition of self-interest as a natural and acceptable force. Such attitudes weakened the structure of traditional society. "The medieval doctrine of natural law did not give free play to the self-interest of a group of 'economic men' bargaining in a free market."[211]

Thomas Wilson attacked such self-interest and individualism in his *Discourse Upon Usury*. Wilson's values "derived ultimately from religious sources."[212] Yet he was aware of the attitudes that separated economic purposes from religious ones, and his literary merchant, "Gromelgayner," represented and expressed these secular ideals. "Let the worlde care for it selfe, and let everyone aunswer for his own doings."[213] Understatement was not Gromelgayner's style.

Gromelgayner appreciated the values that motivated individual economic activity. In fact, he understood the positive value of an active life. "I have bene a doer in this worlde these 30 winters."[214] According to him, there could be no expectation of commerce if the merchant could not aspire to gain. "None will lend for moon shine in the water; and therefore, if you forbid gaine, you destroy entercourse of marchandize, you overthrowe bargaininge, and you bring all tradinge betwixte man and man to such confusion, as . . . man wil not deale."[215] The public welfare depended upon a healthy foreign commerce, he argued, but "where no gayne ys to be had, men will not take paynes."[216] Consequently, he maintained that "merchants doings must not be overthwarted by preachers and others, that can not skill of their dealings."[217]

Of course, Wilson's merchant defended his own life-style very crudely. Eventually, persuaded by a preacher of his sins, he recanted. During the dialogue, however, a civil lawyer introduced a middle view which allowed that usury had practical advantages for society but that it must be condemned when it "bites."[218] Wilson's dialogue, then, dramatized a range of views, each representing some shade of contemporary opinion about private economic purposes in general and about usury in particular, and each ultimately irreconcilable with the other opinions.

The shift in attitude from a traditional to a secular understanding of economic activity was featured in the profusion of pamphlets that debated about the propriety of usury. The three general opinions outlined in Wilson's pamphlet were reiterated time and again. No other form of economic behavior seemed to threaten traditional views as did the taking of usury. In simple terms that harked back to order theory, usury blasphemed God, and rejected the fullness of his creation. Robert Mason wrote that a usurer was "not contented with the things of Gods creation, according to their own values," he wished to "adde further provisions and means than ever was ordained of God. Out of these reasons," Mason concluded, "the raysing of increase upon bare money, is unnatural . . . because God hath left nothing unprovided that is requisite for the use of man."[219]

For these reasons, and since it "appeared to produce no tangible commodity, yet it was labor that never rested,"[220] usury indicated the more generalized redefinition of the idea of order and identity during this time. And "the figure of the usurer was as much a rhetorical touchstone for the debate over a secular economy as the figure of Machiavelli had been in the debate over a secular polity. [He was] an 'artificial person' . . . like the actor, a luminary: a marginal man perennially poised in midpassage,"[221] a figure situated like Macbeth or Coriolanus, on the borderline, suggesting simultaneously the aggrandizing and the nothing "self."

Traditional theorists dwelled upon the unnaturalness and the ungodliness of usury, and managed also to establish a causal relationship between such usury and social chaos. Miles Mosse argued that lending was free in nature, and that to couple lending with gain created an unnatural beast.[222] He blamed usury for nothing short of the demise of commonwealths, and rhetorically explained that the confusion and destruction caused by usury was the result of the "want of due proportion committed in the very manner of the contracting itself."[223] Finally he looked to history and uncovered the wisdom of the "great common-wealth men of all ages," who prohibited usury because "it and the publique good, could not well stand together."[224] Mason concluded likewise, and he too blamed unnatural usury for a spate of social ills: "Usury is set up as an universal Trade. This is the cause, that charitie groweth cold, loving affection between friends and alies turned into hatred . . . and the service of God despised."[225]

Yet the commercial spirit that began to characterize English values during these years even influenced as traditional a minister as was Miles Mosse.[226] While biblical argument permeated his treatise, Mosse still attempted to legitimate some forms of lending; indeed, he differentiated, in a crude way, between moneylending and investment banking.[227] Others more outrightly accepted the taking of usury. Thomas Culpeper admitted that religion had little control over men's consciences in his day. He accepted usury as commonplace economic activity, but he sought legislative redress for high rates because he believed they inhibited trade.[228] Thomas Mun based his support of usury on economic foundations. Unconcerned with religious issues, Mun believed that usury facilitated the circulation of money which was requisite for successful trade.[229]

Gradually the religious accretions that weighed down the discussions of economic activity and of usury fell away and theorists argued these topics on economic grounds alone. Bacon understood that usury was rather a practical concern than a religious issue. "Since there must be borrowing and lending, and men are so hard of heart as they will not lend freely, usury must be permitted."[230] With a Hobbesian proclivity to psychologize, Bacon concluded that it was "vanity" to expect that money would be lent without profit.[231] In balancing the advantages and the disadvantages of usury, Bacon considered economic consequences alone. Usury might limit the flow of money, lower the price of land, diminish the number of active merchants, decay the customs of monarchs which were related to merchandising, and consequently inhibit industry and manufacturing which relied upon the steady circulation of money, he thought.[232] Conversely, he argued that borrowed money could be used to increase trade and to enable men to maintain their property.[233] As did others, Bacon distinguished between harsh rates of interest and moderate rates. He felt, too, that the government should legislate that landowners and the general public could borrow money at 5 percent and that merchants could borrow money from special licensed lenders at a greater rate. This, he believed, would insure that money would be available for commerce and trade, where otherwise merchants might have trouble getting loans. And although, he admitted, such legislation clearly authorized usury, he concluded that it was "better to mitigate usury by declaration than to suffer it to rage by connivance."[234]

Around 1630, Robert Filmer, the divine right theorist and author of *Patriarcha*, wrote a pamphlet that examined and refuted an earlier treatise by Roger Fenton which declared that usury was unnatural. While unable to transcend biblical and religious argument totally, as Mun and other merchants did, Filmer chose to prove that the Scriptures did not proscribe usury and that because of this, usury was not "unnatural, ungodly, or uncharitable."[235] Indeed, in the preface to the pamphlet, he established the general argument that usury was a matter of civil law alone. He would show, he claimed, that "usury is no where in Scripture forbidden to Christians: but that it is as lawful as any other contract or Bargain, unless the lawes of the Land do prohibit or moderate it, as a point of state or policy."[236] Additionally, he argued that usury was, likewise, never forbidden to Jews, except in one special "politick" law of Moses which was conditional to particular circumstance.[237]

Filmer's opinions on this subject evidenced how directly shifting economic values influenced general secular and political concerns during these years. His claim that the biblical law regarding the Jews and usury reflected special historical circumstances rather than natural laws, was important. From this, he argued that "we are left to the laws and customs of the Kingdom to guide us in our Contracts."[238] And with practical insight into the temper of his times he realized that usury was "the foundation and rule which governs the valuation of all other sorts of Bargains."[239]

Once he established his legal case for usury, Filmer continued to evaluate the practical and the realistic necessity for its existence. After considering civil law, he explained that "it is fit therefore in equity, that since the Lender stands in hazard there should be a gain due him also."[240] As did others, he too defended moderate rates, but he based his view on the empirical nature of contractual definition. He stated that

> it is rarely to be shewed that any loss can befall a man in life [and] goods merely by the act of God, without the concurrence of some fault of man, either of negligence, ignorance, indiscretion, willfulness, or the like. . . . the rule that guides the valuation of all Contracts, is not what *Casually is or may be*, but what *ordinarily* is like to happen.[241]

Filmer's discussion of usury, then, inspired him to practical and somewhat empirical economic and legal analyses of contemporary human relations, and fostered in him a developing awareness of the burden of individual responsibility which elsewhere he described as the patriarchal nature of order.[242]

Another explicit change in attitudes dramatically detailed in the mercantile pamphlets concerned merchant companies and free trade. Early in the century John Wheeler, even while he sought ways to improve private gain and public profit, advocated merchant companies and favored ordered and regulated trade.[243] Such conviction represented privileged position as well as a lasting affinity for the idea of order. Such men as John Roberts and Lewes Roberts, however, expressed the opposite opinion. In *The Treasure of Traffike*, Lewes Roberts declared that men should "not suffer any Monopolies, Pattents, and grants to private men, which may hinder the liberty and freedom of Traffike."[244] John Roberts favored free trade also. It increased the king's customs, he maintained, and it benefited private merchants because they could trade freely anywhere they pleased. Most important, however, free trade made

> the universal body of the subjects of the land content, in that they may become merchants, being very ready . . . to make discoveries, whereas now otherwise merchandize sorting [and] setled in companies, confineth merchants into those limits that private orders tie them in, so they may not helpe themselves through any discouragements in one trade.[245]

Thomas Mun described the qualified merchant. He listed characteristics that indicated a new understanding of economic activity. Besides knowing languages, arithmetic, and navigation, the merchant had to understand accounting, customs and taxes, bills of exchange, and the consumption requirements of foreign nations.[246] With Mun, "for the first time economic factors were clearly differentiated from their social and political entanglements."[247] The secular attitudes, advocated and advanced by Mun, Robinson, L. Roberts, and others in their attempts to explain and justify economic activity, became fundamental to a redefined social order founded on possessive market values and sovereign political representation. "This secularization of economic life was basic to the triumph of a capitalistic organization grounded upon the right and efficacy of private, economic decision making. The ideology that made possible the material orientation of modern society had to facilitate a shift of loyalties from the sacred to the profane without a disruptive demoralization of society."[248] It seems fair to suggest that the merchant pamphleteers played an important role in the production and dissemination of these new values. In attempting to prove their social worth they generated a secular discussion of economic activity and provided a socially responsible example of the expression of self-conscious interest and identity.

The appreciation of man's capacity to affect social change, to define social utility, and thereby to secure order in a changing world was not limited to a self-conscious merchant citizenry or to political theorists such as Sir John Eliot. Developments in a wide range of religious, artistic, and philosophical thought approximated the new political and economic understandings of the relationship between self and society, and to a large degree served as conservative cosmological foundation for the innovative concepts and activities of the politicians and the merchants. Indeed, characterized by an appreciation of the power of man as "doer," a new understanding of the power and limits of knowledge, and a positive view of the natural world, such intellectual activity helped to celebrate a dynamically responsible sense of self while at the same time it rhetorically described a naturally ordered social world.

Two closely related intellectual traditions positively influenced the secularization of social meaning in early modern England and added to the innovative as well as to the generally conservative character of this process. Although the Hermetic-Rosicrucian and the developing "modern science" movements ultimately went separate and inimical ways, during these years their many common characteristics and concerns often linked them.

Both were interested in man's will. Man as "operator," whether as the magus, as Faustus, as Prospero, or as the socially concerned scientist member of the Royal Society, captured the English imagination during these years. Both traditions investigated the world; the search for knowledge and the concern with the power of knowledge were ubiquitous activities for them. The Hermeticists, as well as the "scientists," stressed the use of numbers and of mathematics and both valued social utility.

Perhaps the most conspicuous intellectual link between the two traditions, however, was Francis Bacon. It is in his philosophical tracts, in his essays on learning and knowledge, and especially, in his *New Atlantis* that the close relationship between these traditions becomes apparent. But these same works clearly reveal the more significant relationship between these movements and the new secular and practical values concerning self and society which helped redefine social meaning and the idea of order.

It is as an alternative means to an ordered and certain existence which influenced some of the most sensitive and intellectual men of the late sixteenth- and early seventeenth-centuries, men such as Sidney, Hooker, Hayward, Forset, Sir John Eliot, Shakespeare, and Francis Bacon, that the Hermetic tradition should be considered here.[249] Changing conceptions about nature, knowledge, magic, and science played a part in the profound emphasis on social utility as much as new political and economic ideas did, and provided some theorists with the "means" with which to confront a social world they perceived as decaying and an emotional world they characterized as evanescent.

During the past twenty years scholars have begun to describe the connections between the Hermetic tradition and religion, science, and magic. The goal of the Renaissance magus, influenced by Hermetic teachings, was to control nature. Hermeticism stressed the conviction of man's divine creative possibilities; man could manipulate and direct nature.[250] It was "a way to infuse morality, piety, faith and awe into a worldly religion."[251] Peter French understood that through this tradition, magic and practical application combined to transmit the knowledge of the age.[252] It served as a bridge that facilitated confidence in man and knowledge. John Dee, a magus and the "prince of mathematicians," conceived of a cosmos that resembled the three-sphere cosmos of Pico and Ficino, but which borrowed heavily from Pythagoras and from the German magician Agrippa for much of its functional definition. Mathematics was the key to this triadic universe.[253]

The literature known as "Hermetica" had two applications. Those parts that dealt with alchemy, magic, and astrology stressed operation in the world; the philosophical tracts expressed a contemplative gnostic appreciation of man's divinity.[254] Man, then, could be omniscient and omnipotent. He was capable of understanding and of controlling the world.[255] Surely Marlowe's characters were the secular counterparts of these divinely inspired and Hermetically influenced men.

Hermetic religious philosophy shaped the operational aspects of the magical tradition and directed this activity toward practical purposes. The high opinion of man and of his possibilities and the mathematical quality of operational magic became interrelated characteristics of the new interests in the study of nature and in social utility. "Hermetic texts encouraged a basic psychological change that released the human spirit and thus promoted . . . [men] to experiment with the powers of the universe.[256] Indeed, John Dee proposed an operable theory of experimental science well before Francis Bacon did.[257]

The emphatic change that this tradition wrought concerned the will of man. The cosmos that these magicians and philosophers worked in and contemplated resembled that of the traditional "World Picture."[258] Certainly, such philosophical conceptualizing confused the traditional correspondences and sympathies, but generally the activities fostered by this tradition, while vital and innovative, were experienced in a known cosmos. Nonetheless, as Yates clearly argues, it was during these years that theory and practice began to approximate each other. And the magus' desire to operate "changed the will" of man, and dignified activity so that social practice and utility became acceptable.[259]

The Hermetic-Rosicrucian tradition which Yates believes flourished at least into the third decade of the seventeenth century directed its religious-philosophical tenets toward social reform. An Eirenicist vision in response to the religious dissension of sixteenth-century Europe might have inspired this reforming consciousness.[260] Bruno and seventeenth-century theorists identified with this tradition emphasized public service, learning, and social utility as the foundations for a reformed society.[261] Indeed, the association of learning and education with "Illumination" in much of the early seventeenth-century Rosicrucian literature has led a few theorists to tie Bacon and his reforming zeal to this tradition.[262] Still, man could activate social reform through self-conscious religious and educational reform; this was understood "as a necessary accompaniment of the new science."[263]

While Bacon's work on knowledge and nature provides the most sophisticated philosophical expression of the new concern with social utility and with social reform, other late sixteenth- and early seventeenth-century literatures illustrate particular themes common to this concern; and Hermetic influence is evident in many of them. Before looking to Bacon, then, for a final evaluation of the place of nature, knowledge, and magic in the new idea of social order and social utility, it might be advisable to discuss briefly some of these other expressions.[264]

Orphic magic was incorporated into the Hermetic tradition just as alchemy and Cabala were. This aural operation could be expressed through music and voice. Poetry, too, might be understood in this way. French claims that Sidney's *The Defence of Posey* recognized the mysterious and magical quality of harmony and music.[265] Sidney's comments about poetry and music remind French of Dee's discussion of music in "Mathematicall Preface," and he maintains that they both followed the same tract for their ideas—traceable to Ficino.[266] Sidney's references to Agrippa in the *Defence* reveal at least a passing knowledge of the Renaissance magic tradition, and French suggests that Sidney's "Poet" is much like the magus. The *Defence* and the basic concepts of Hermeticism stress man's knowledge of himself and man's reuniting with divine knowledge.[267] Indeed, Sidney referred to the poet as the "first lightgiver to ignorance," and venerated ancient knowledge which he linked with the music poetry of Homer, Hesiod, Amphion, and Orpheus.[268]

Still, as French rightly explains, for Sidney, the aim of poetry was "not *gnosis* but *praxis*."[269] Poetry should be didactic, Sidney argued; and he hoped that it could actively serve English political and religious concerns.[270] Indeed, Sidney's friend and subsequent biographer, Fulke Greville, made reference to Sidney's desire for "a general league in Religion . . . an uniform bond of conscience, for the protection . . . of Religion and Liberty"[271]—a reference that gives some credence, perhaps, to Yates' and French's suggestion that Sidney was involved in an Eirenicist league with Bruno. Nonetheless, Greville's Sidney emerges as the representative of a lost golden age, an age that Greville attempts to reclaim, "even in our minds"; an age that he compares to the early seventeenth-century by illuminating the differences "between the reall, and large complexions of those active times, and the narrow *salves* of this effeminate age."[272] Of Sidney, he remembered, "the truth is: his end was not writing . . . but in life and action, good and great."[273]

In his *Life of Sidney*, Greville created an imaginary, ideal world against which he could compare and berate the present, real, and "naturally imperfect" world. Perhaps with a more depressing sense of mortality than the late Shakespeare and the late Bacon, but with a similar melancholic self-consciousness, he looked to an idyllic past to escape "naturall vicissitudes" and "changes in life," and he rather sought "comfortable ease of imployment in the safe memory of dead men, than disquiet in a doubtful conversation amongst the living."[274] As are the imaginary *New Atlantis* and Prospero's island, Sidney's England is a world characterized by a blend of active and contemplative, of variety and unity; and Greville's Sidney embodies these qualities just as surely as Prospero does. Sidney serves his queen and his religion as the active hero, but he retains a sense of an inner private self: poetic, spiritual, and contemplative. And to balance Sidney's spiritual, contemplative person, Greville creates the fully active Essex[275] whose less perfect public life is like the dead Sidney's shadow.

With Shakespeare's late plays, Bacon's *New Atlantis*, and a spate of other early seventeenth-century English art and literature, the *Life of Sidney* treats the interrelated themes of a golden age, of death, and of the conflating of actives and passives. Greville's prose in the *Life*, just as Shakespeare's poetry in *The Tempest*, formally advanced the theme of the unity of actives and passives. Phrases such as "captivated and captivating appearances," "support or be supported," and "contented and a contenting soveraigne," are characteristic of the work.[276] The magical and alchemical quality to this sense of unity was well documented during these years. Divinity had most perfectly accomplished this unity in nature and the neoplatonic-Hermetic tradition taught that man could imitate divinity and could attain the knowledge of this unity in love or by properly understanding nature and operating in it. Donne's "The Anniversaries" captured the poetic and the mystical sense of the relationship between creativity, knowledge, and unity.[277] And in the *Sermons*, Donne clearly articulated the divine, alchemical possibilities therein:

> Love is a Possessory affection, it delivers over him that loves into the possession of that he loves; it is a transmutatory affection, it changes him that loves, into the very nature of that he loves, and he is nothing else.[278]

The lover becomes love in the loved just as the creator becomes creation in the created and as man becomes divine in knowledge. The alchemy, here, is very special.

Agrippa's straightforward explanation of how the magician explores, serves, and interprets nature, which was translated into English by John Stanford in 1569, influenced much of the serious consideration of man's relation to nature during the following years. In *Of the Vanitie and Uncertaintie of Artes and Sciences*, he claimed

> magicians are like careful explorers of nature only directing what nature has formerly prepared, uniting *actives to passives* and often succeeding in anticipating results so that these things are popularly held to be miracles when they are really no more than anticipations of natural operations.[279]

During the Middle Ages magic was often confused with necromancy. Now with man's place and meaning in the world and his relation to that world undergoing redefinition, magic was beginning to be accepted as "a 'human science' worthy of mankind."[280] As Rossi clearly noted, "magic ceased to be the disturber of universal order and of a fixed celestial structure when order and structure began to be contested from all sides."[281] Bacon appreciated that magic aided man's understanding of nature. His definition of magic owed much to Agrippa, as is clear in the following:

> [Magic is] the science which applies the knowledge of hidden forms to the production of wonderful operations; and by *uniting (as they say) actives with passives* displays the wonderful works of nature.[282]

Greville's *Life of Sidney* and much of his poetry as well as a great deal of other early seventeenth-century art and literature shared with Raleigh's works a haunting sense of cycles, decay, and impending doom. A morbid interest in death, conspicuous in the *Life of Sidney*, captivated the artistic consciousness of these years. But this haunting vision found expression in another way, too: in the already-mentioned theme of the golden age. Greville brought to this vision his overly melancholic despair and wallowed in an ideal past that he claimed was surely gone. Others, however, used the sense of a golden age to help exorcise the fear of change and mortality. Thomas Acheley, for instance, imagined that Elizabethan England witnessed a resurrected golden age; history had turned full cycle and virtue was reborn in his own time: "England thou are Pleasures-presenting stage, / The perfect patterne of the golden age."[283]

Regeneration was the key to the positive character of this theme, especially as it was dramatized at James' court during the first decade of the century and in Shakespeare's last plays. The masques that Ben Jonson and Inigo Jones, among others, directed for the Jacobean court celebrated the pious and imperial splendor of the court and created an idyllic historical meaning for the monarchy which recaptured the memory not only of the Elizabethan age but of the mythic and antique English past as well. As Yates has argued, the religious aspects of the Hermetic-Rosicrucian tradition colored this resurrection theme in court entertainments as well as in Shakespeare's late comedies.[284] Renewal and regeneration promised peace, true religion, and social well-being. As a means to confront change self-consciously, this religiously inspired theme evaded historical responsibility for order and social reform but helped Englishmen prepare for that responsibility by ridding them of their hopeless fear of change and inevitable decay. In celebration of James and of his daughter Elizabeth's marriage in 1613, Shakespeare invoked this theme in Henry *VIII* more powerfully than he had in any previous work. From Elizabeth (Tudor) to Elizabeth (Stuart) shall pass "a thousand thousand blessings," Cranmer augurs to Henry after Ann gives birth. Elizabeth I will not perish, but like the phoenix, she will be resurrected:

> Her ashes new create another heir
> As great in admiration as herself,
> So shall she leave her blessedness to one
> (When heaven shall call her from this cloud of darkness)
> Who from the sacred ashes of her honor
> Shall star-like rise, as great in fame as she was,
> And so stand fixed. Peace, plenty, love, truth, terror,

That were the servants to this chosen infant,
Shall then be his, and like a vine grow to him.
Wherever the bright sun of heaven shall shine,
His honor and the greatness of his name
Shall be, and make new nations. He shall flourish,
And like a mountain cedar reach his branches
To all the plains about him. Our children's children
Shall see this, and bless heaven.[285]

Perhaps it was just that he was getting old, but nonetheless even the great Shakespeare confronted the change and uncertainty of history by creating for it an idyllic meaning. Yet while the vision was one of resurrection the meaning was not finally religious; the present manifested a historical, imperial and ultimately, then, a secular purpose. And such purpose could only continue to be expressed in the historical future. It was not, then, such an impossible task for Francis Bacon to create a vehicle that would carry him to an island in the historical present whose golden age resulted from, among other things, the benefits of social reform and social utility.[286]

Before he discovered Bensalem, however, Bacon had already discovered the worth and power of knowledge as an agent for social reform and utility, and he believed that true knowledge alone could indicate social meaning. The "true ends of knowledge," he claimed, are not for pleasure, fame, power, or profit, but for "the benefit and use of life."[287] And toward such ends he sought not "things in accordance with principles, but . . . principles themselves."[288] Upon these two general convictions, Bacon established a "meaningful" understanding of the relationship between man and nature. The knowledge he sought and, subsequently, the knowledge he sought to use for social ends ordered man in nature and indicated Bacon's practical appreciation of social and political reality.

Even as a young man, Bacon denigrated traditional knowledge for being barren of any useful manifestations or inventions.[289] In so doing, he placed himself among a growing number of Europeans, many of whom expressed Hermetic and neoplatonic attitudes, who vehemently wished to replace traditional knowledge with a practical and operational knowledge of nature. As Rossi convincingly stated,

> Bacon was voicing the general opinion of his age, defining some of its essential demands, when he strove to rehabilitate the mechanical arts, denounced the sterility of Scholastic logic, and planned a history of arts and sciences to serve as foundation for the reform of knowledge and of the very existence of mankind.[290]

The reform of knowledge, then, would be the effective cause of the reform of society. And even for Bacon, reform meant return and resurrection. In a manner that looked back to the gnostic tradition of Joachim of Fiore and foreshadowed the romantic posturings of Friedrich Schiller and Heinrich von Kleist, for example, Bacon expected that a renewed state of grace for man in nature would accompany successful social reform. The separation of man from nature through false conceptions of the mind conditioned and characterized "fallen" society. Correct knowledge was grace. And toward such eschatological ends Bacon maintained that "all trial should be made, whether that commerce between the mind of man and the nature of things . . . might by any means be restored to its perfect and original condition."[291]

The religious implications and imagery of such a "reformation" notwithstanding, Bacon's plans for social reform were inevitably time-bound. Man had to create the necessary conditions for a reformed knowledge, and consequently for reformed society. This required a practical, empirical appreciation of social reality and an active participation in "history." If Bacon did not see himself creating new knowledge, he thought that he could prepare the way for the reformation of proper knowledge. In this fashion, he would be actively participating in nature; but he would be following nature. As he wrote to Burghley in 1591:

> I have taken all knowledge to be my province; and if I could purge it of two sorts of rovers, where of the one with frivolous disputations, confutations, and verbosities, the other with blind experiments and auricular traditions and impostures, hath committed so many spoils, I hope I should bring in industrious observations, grounded conclusions, and profitable inventions and discoveries; the best state of that province.[292]

So, by exorcising knowledge of the purposeless intellectual conjurings of the scholastics and of the thoughtless operations of the conjurors, Bacon would prepare the way for a knowledge that married thought to nature and consequently benefited society.

The ability to tap the reforming power of knowledge depended on the method used to gain knowledge. The proper inquiry into nature reformed man and emancipated him from the limits of his perspective. He was then capable of redefining society according to its natural potential. Man's reasoning faculty had to be framed, shaped, and fed with correct information about the world. Bacon's "reason" in no way resembled "right reason" whose innate character reflected and participated in nature's order. Man must construct an edifice of reasoned observation because "the entire fabric of human reason which we employ in the inquisition of nature, is badly put together and built up, and like some magnificent structure without any foundation."[293]

Even though his own thought depended to such a large degree on essentialist construction, Bacon belabored the importance of observation in directing or guiding thought. Because of this he ultimately rejected even his teachers, the "more ancient of the Greeks," who trusted "entirely to the force of their understanding, applied no rule, but made everything turn upon hard thinking and perpetual working and exercise of the mind."[294] Yet the meaning and the power of knowledge were not to be found in empiricism alone, either. Bacon was no positivist, and his world was not yet ordered by methodological inquiry. But, by careful observation, man could participate in nature's *essential* order and, in so doing, resurrect the natural union of mind and nature. As he lamented,

> while men are occupied in admiring and applauding the false powers of the mind, they pass by and throw away those true powers, which if it be supplied with the proper aids and can itself be content to wait upon nature instead of vainly affecting to overrule her, are within its reach.[295]

Like the magus, Bacon appreciated the union of aspiration and order. Methodologically, he suggested a means to direct man's potential responsibility for order toward temporal ends and, at least for the immediate future, resisted the existential void which Marlowe's characters had threatened to enter.

Bacon, then, still believed in a natural order, and he felt that man could conceptualize this order; he was not prepared, as was Hobbes, to rid order of any predefined eschatological essence. But his understanding of knowledge anchored the conceptualization of order to a methodology based upon an empirical appreciation of social reality. And with this empirically essential methodology he built a bridge from which he could easily look to the future shore. His method for knowing clearly resembled Hobbes':

> Those however who aspire not to guess and divine, but to discover and know; who propose not to devise mimic and fabulous worlds of their own, but to examine and dissect the nature of this very world itself: must go to the facts themselves for everything.[296]

Man's mind, then, could reflect social reality, and it could know order, but false worlds and assumptions had first to be destroyed. Only then would the "wedlock of Mind and Universe" recur.[297] Bacon complained that men constantly tried to fit all variety of things into a fixed order and to define all things by correspondence. "There are many-thyngs in nature as it were *monodica, suijuris,* (singular, and like nothing but themselves;) yet the cogitations of man do feign unto them relatives, parallels, and conjugates, whereas no such thing is."[298] And along with the traditional "Order theorists," he took the Renaissance magicians and Hermeticists to task as well. Paracelsus and the alchemists and their followers have grievously strained the ancient opinion that man was a microcosm by exaggerating the correspondences and parallels with the macrocosm, he maintained.[299]

For Bacon, knowledge opened man's eyes to nature. As did so many of his contemporaries, he continually struggled with questions of permanence and change; and his ordered universe contained eternal matter which was ever in flux.[300] He neither sought nor celebrated first causes. "Order . . . was for him not a moral desideratum at all but a simple guarantee of the existence of unchanging physical laws."[301] Both the knowledge of man and the knowledge of nature enlightened man about efficient causes and both had ultimately to be defined "in terms of power and utility."[302] As Bacon aphoristically stated: "Towards the effecting of works, all that man can do is put together or put asunder natural bodies. The rest is done by nature working within."[303] This was an important insight because it allowed man, the manipulator, the freedom to shape things; but it recognized that what was to be shaped already existed. Bacon, then, conservatively rejected traditional notions of nature's autonomy over order and social reality, and prefigured the new concepts concerning nature, society, and order which Hobbes and others would soon articulate. Bacon extolled man the mechanic, the actor, and downplayed man the thinker. His understanding of nature and of man helped define a new idea of order which no longer described a divine cosmos corresponding to man's little self; rather, order was social reality itself, shaped and framed by human will and action.

Bacon worked out his understanding of nature and order in *De Sapientia Veterum*. As he looked to the wisdom of ancient fables for allegorical truths about political and social reality, so too he sought in these myths knowledge about man's relation to nature. And in "Pan; or Nature," and "Orpheus; or Philosophy" he described a full, but changing universe which served as the basis for his subsequent understanding of man's responsibility for order and social reform.

Immediately Bacon manifested an indifference to precision concerning first causes. The world originated from One or from Many. Oneness was potentially many, anyway.[304] And Pan, or nature, was potentially Pan-ic. Confusion, anxiety, and fear defined all nature, including man. Pan challenged "Cupid; or The Atom," the "instinct of primal matter."[305] Bacon recognized in this potential an inclination of all matter "to dissolve the world and fall back into the ancient chaos."[306] Only the concord of things, love as order, could mitigate such propensity. Variety, the potential many (Lovejoy's "Goodness") characterized nature, but one of its parts, "Man," had to frame and control it if chaos was to be resisted.[307]

In a typical Renaissance cosmological fashion Pan needed no lovers because "Plenitude" defined nature. Yet Pan had a wife, Echo. Reality, then, manifested itself. It was not represented anywhere, but it could be repeated (echoed) by true knowledge (Philosophy).[308] Finally, Bacon's understanding of a full, material, and changing universe found allegorical signification in the fact that Pan had no offspring. Generation occurred within the cosmos, but the whole was full and finite.

Bacon struggled with the sense that both order and disorder were natural. Since it was impossible to retard "dissolution and putrefacation," Bacon argued, natural philosophy turned to human affairs and conditioned the ideas of harmony, virtue, and peace in man.[309] Therein was the origin of civil order because natural philosophy "teaches the peoples to assemble and unite and take upon them the yoke of laws and submit to authority, and forget their ungoverned appetites."[310] With Hobbes, Bacon understood that order was necessary. But for Bacon, nature created a conceptual order; for Hobbes, man contrived a material order for himself. Bacon's order was not definitive, rather it participated in the natural flux of things. Since nature meant order and disorder, Bacon claimed, civilization flourishes; then seditions and rebellions break out and finally "according to the appointed vicissitude of things," civilization again arises "not in the places where they were before."[311]

Bacon seemed to have fallen prey to the same sense of doom and cosmic determinism that haunted Raleigh. Nature's course threatened man's will and his actions as surely as God's providence had in traditional order theory. But a rhetorical distinction of the utmost importance created a way out for Bacon and for Bacon's mankind. Bacon construed natural order conceptually and natural disorder materially; and man's natural material imperfection, his mortality, motivated him to take advantage of the gap between these two natures. It was this imperfection that set "men upon seeking immortality by merit and renown."[312] Bacon sensed what Marlowe, Shakespeare, and Hobbes felt: man's mortality spurred him to act, to improve, to create. So within the eternal, cosmically determined frame of things, man acted, and defined meaning, in time.

The New Atlantis described a real historical society which had survived in its present condition for three thousand years.[313] With the aid of knowledge and religious unity, the citizens of Bensalem had reformed their world and had seemingly transcended the web of historical cycles and fortune. But for all its apparent realistic description, *The New Atlantis* was an allegory, and Bensalem was a fabulous world. Early in their visit the Europeans sensed the mystery that surrounded the island. Their leader noted that they were beyond the Old World and the New, between life and death. They had no identity; their tradition and their history meant nothing here. They were between light and dark. They were in a fable.[314]

Much of *The New Atlantis* is acroamatic writing.[315] Just as he looked to ancient fables for wisdom about nature and political reality, so, late in his life, Bacon created his own allegorical tale to teach his world about the possibilities of social reform. The story served as a focal point for Bacon's ideas about knowledge, nature, magic, and power. As had always been Bacon's definitive purpose, *The New Atlantis* would serve mankind.

Others have pointed out that *The New Atlantis* reflected Hermetic and Rosicrucian concerns and sympathies. The emissary from the island presented the Europeans with a scroll written in ancient Hebrew and ancient Greek and signed with "a stamp of cherubins wings, not spread but hanging downwards, and by them a cross."[316] The officials of the island wore white turbans emblazoned with red crosses. Francis Yates carefully detailed the affinity between these symbols in *The New Atlantis* and the Rosicrucian literature published early in the seventeenth century.[317] The Bensalemites expressed a reforming zeal similar to that found in Rosicrucian and Hermetic ideas. The official priests of the island were also healers of society. Yates concluded that Bensalem was governed by Rosicrucians.[318]

The crew noted the mysterious ambience, too: "There was something supernatural in this island; but yet rather as angelical than magical."[319] The island guarded its knowledge as if it were a treasured secret. The "Merchants of Light" traveled around the world, incognito, so "we know well most part of the habitable world and are ourselves unknown."[320] Their purpose, as befit their name, was to acquire the commodities of knowledge: books and experiments.[321] Religion, too, came to Bensalem as knowledge, in a sign of a cross of "light" rising from the ocean during the night.[322]

New Atlantis embodied the visionary ideal of unity that had characterized the reforming purpose of much Hermetically and neoplatonically influenced thought and action since, at least, Pico's time. Bensalem was sect-less. Christian union (the goal of Sidney and others) prevailed. Difficulty of communication was bypassed when an Ark with God's word was found and each who looked at it "read upon the Book and Letter, as if they had been written in his own language."[323] Bacon even confounded the usual geographical distinctions of east and west in *The New Atlantis*, adding a spatial and cosmological dimension to the religious unity in this fable.[324]

But the capacity to discern true knowledge, and the ability to use knowledge for the benefit of man and society, formed the structural and the functional foundation of *The New Atlantis*. The Bensalemites appreciated both the natural and the mysterious relationship between knowledge and creation. One of their wise men accepted the pillar of light from God and articulated a profound understanding of this relationship:

> Lord God of heaven and earth, thou hast vouchsafed of thy grace to those of our order, to know thy works of creation and the secrets of them; and to discern . . . between divine miracles, works of nature, works of art, and impostures and illusions of all sorts.[325]

Since natural laws defined so much of social reality, he knew, miracles would rarely occur. As Bacon had so often taught, the key to knowledge and to the control of knowledge resided in a proper application of a discriminating methodology.

The order of "Salomon's House," "dedicated to the study of the Works and Creatures of God," controlled the knowledge aggrandized by Bensalem's "Merchants of Light" and translated it into practical inventions that benefited society.[326] As one of the Chief Brethren explained: "The End of our Foundation is the knowledge of Causes, and secret motions of things; and the enlarging of the bounds of Human Empire, to the effecting of all things possible."[327]

The New Atlantis celebrated the magical, the religious, and the secular aspirations of mankind and united them around the goal of social utility and reform. Only true knowledge could act as the refining force and the controlling agent for the "effecting of all things possible." "If there be a mirror in the world worthy to hold men's eyes, it is that country," remarked Bacon's narrator.[328] A mirror as a reflection of God's cosmos and nature's order was a commonplace concept in traditional correspondence theory. Bacon doubted the clarity of the traditional "mirrors"; he denied, too, that man himself was a worthy reflection of nature's laws. But *The New Atlantis* reflected the highest possibility that man could aspire to: redeemed and reformed social reality.

CHAPTER 5

The Final Defining Voices

> I now come to a Conclusion, and I have nothing to propound to your Lordships by way of Request. . . . The Commons will be glad to have helpe and concurrence in saving of the Kingdome; but if they should fail of it, it should not discourage them in doing their dutie.
>
> John Pym, *A Speech delivered at a Conference of Lords & Commons* (January, 1641)

> The question is not, whether there shall be an arbitrary power; but the only point is, who shall have that arbitrary power, whether one man or many.
>
> Sir Robert Filmer, "The Anarchy of A Limited or Mixed Monarchy or A succinct Examination of the Fundamentals of Monarchy"

The year 1652 witnessed the first printed critique of Thomas Hobbes' *Leviathan.* Sir Robert Filmer, the author of the review, had been contemplating and writing about the origins of government for at least sixteen years; much of what he had concluded about sovereignty, arbitrary power, and law might have predisposed him to look with favor upon Hobbes' masterpiece. Indeed, Filmer found areas of agreement between Hobbes' ideas and his own, but a major difference overshadowed these agreements. As Filmer stated: "I consent with him about the rights of exercising government, but I cannot agree to his means of acquiring it."[1] This was crucial. Over the previous two decades, a host of politicians, polemicists, and theorists had articulated the utilitarian benefits of a socially constructed order, and suggested that a variety of governments could, indeed, exercise that order. But Hobbes' work definitively divorced the efficient exercising of government from the question of the acquisition of power. More specifically, Hobbes emancipated the secular state from society and instituted its sovereignty as social order.

Filmer's critique is an integral part of that social product which textually redefined social meaning and which Hobbes translated into objective reality. Hobbes eschewed ethical and moral considerations about the acquisition of power. He rid the state and civil order of eschatological purpose. Filmer could not. Even in 1652, Filmer sought beyond political reality for an essential nature to order and government.

Hobbes deduced political theory from political reality and from his understanding of man's motive nature. Filmer shared little of Hobbes' psychological insight. If all men were equal in a state of nature, asked Filmer, why should war exist?

> God was no such niggard in the creation, and there being plenty of sustenance and room for all men, there is no cause or use of war till men be hindered in the preservation of life, so that there is no absolute necessity of war in the state of pure nature; it is the right of nature for every man to live in peace.[2]

Filmer did not identify human invention as the source of either change or order in the world. While he acknowledged the ultimate authority of civil order and recognized, too, that flux and change were natural, he rejected the idea that men contrived or were responsible for any of this. In his famous *Patriarcha*, written in the late 1630s, he argued that nations developed as sons naturally accepted relegated power from the sovereign father.[3] By marrying sovereignty and order to natural paternity, a position that Locke was to challenge directly later in the century, Filmer resisted the burden of responsibility that others were beginning to assume for change and for order. As the origin of government, paternity insured that social organization, the family, ordered the state. There was no place here for the psychological perspective that understood the state as the contracted order created by frustrated, motivated individuals: sons and brothers.[4]

Political theory and propaganda in the 1630s and the early 1640s dynamically outlined a political philosophy of the secular sovereign state which received its most crystallized articulation in the writings of Thomas Hobbes. Power, policy, and purpose characterized the state which represented man's repressed but creative will. Representation now constituted order metonymically, by substitution rather than metaphorically by correspondence. And authority no longer reflected transcendental order or essential law, but discursively structured and exercised power to create law and order.

Sir Robert Filmer and John Selden dealt with many of these issues in decidedly nonpolemical ways. Each was concerned with political obligation and the origins of political power, and each understood that property formed the foundation for the relationship between obligation and authority.[5] Yet their views were often opposed and their writings, then, provide a good general introduction to the more acrimonious propagandistic pamphlets that discussed these concerns in advocacy of Royalist or Parliamentarian positions in the early 1640s.

Laslett pictured Filmer as a genuinely conservative man struggling to resist social and political change.[6] But wisely he also recognized that, without doubt, Filmer was a "disturbing radical."[7] While he "maintained a deliberate attitude of non-intervention" during the 1640s, Filmer did his rhetorical share in overthrowing the traditional idea of order by tying kingship and sovereignty to patriarchy, property, and obligation.[8] And because of this, he founded his political theories, as he had his ideas about usury, on civil law alone.[9]

Selden, too, must be remembered for his conservative temperament and for his radical impact. While he referred to Parliament as the country's "Supreem Hand," he refused to take sides in the mounting struggle for power and sovereignty which he knew was the prime motive behind most political, constitutional, and religious arguments.[10] Indeed, he denied that those who make the king were greater than the king (p. 93). Men pretend religion in wars to hide their real interests, he noted with insight (p. 121). Those who claim to fight for religion, honor, and truth, he pointed out, usually keep those things in their pockets (p. 138). "Generally to pretend conscience against Law is dangerous," he concluded (p. 35).

Unlike Filmer, Selden looked specifically to custom and common law, "the best Law of the Kingdome" (p. 137), as the final arbitrator in issues social, political, and religious. He defended ceremony, too, because it "keeps up all things" (p. 24). But unlike traditionalists, Selden neither wallowed in correspondence theory, nor tied himself to precedent. Ever practical and inclined to confront reality, Selden, as did Filmer, defended usury because, as he noted: "Tis a vaine thing to say money beggets not money, for that no doubt it does" (p. 135). Similarly, he freed private and political obligations from scriptural precedent. "Abraham paid Tythes to Melchisedek, what then? twas very well done of him, it does not follow therefore that I must pay Tythes, no more than I am bound to imitate any other action of Abraham" (p. 128).

Governments and the obligations created by governments, oaths and laws, are contracts, Selden maintained. Consequently, all jurisdiction is civil (p. 60). Selden refused to consider politics in relation to order theory, divine law, or nature. Government was a practical concern. "Every Law is a Contract between the Prince & the people & therefore to bee kept," he taught (p. 69). The same was true for oaths; yet Selden knew that the king's oath was not security enough because although his oath was governed by law, and judges interpreted the law, "what Judges cann bee made to do we all knowe" (p. 66). For such reasons men must take the responsibility to maintain good government. They must diligently examine the contract upon which their government rested. And they must insure the reciprocal nature of political obligations. To know the nature of personal obedience and of governmental obligation, one must look into the contract, said Selden (p. 137).

On the other hand, the concept of an individual, unconstrained but by his own will, was impossible for Filmer, whose major works directly opposed contractual theories of political obligation.[11] Filmer attacked Parsons (Doleman), Cardinal Bellarmine, and other Catholic political philosophers for advancing the notion that man was naturally endowed, at birth, with the liberty to choose any form of government.[12] And he clearly associated the politics of Calvin and others of the "Geneva discipline" with that of the Catholics: an association, by the way, which conservative theorists and politicians would repeatedly portray for the rest of the century.

Liberty denied the wholeness of the political body. The subjects and the prince, like the father and the son, were of each other, and the property that passed from father to son symbolized this oneness. A radical theorist, who with Hobbes foreshadowed much modern conservative ideology, Filmer resisted the emergence of nominalist politics by reducing all political expression to natural social activity. Meaning was conceptual, to be found in origins, for Filmer; it was not existential.

Man was born obliged to political authority; his father's right of proxy naturally limited his freedom.[13] Political society, then, depended upon personal relationships and the relationship of persons to property. And only private property, family property, existed.[14] By subsuming political activity under patriarchial social functioning, Filmer denied that the state existed separately from social nature. There was no natural state and there had been no state of nature. Political society—social nature—always existed. There could be no question of an arbitrary, contrived, created, or contracted polity.[15] And consequently government did not represent private wills, but reflected patriarchically ordered familial reciprocity. Change occurred, but it was nondefining, and political behavior could not contravene this natural fact. As Filmer argued, "the newcoined distinction into Royalists and Patriots is most unnatural, since the relation between King and people is so great that their well being is reciprocal."[16] Without Hobbes' psychological insight, Filmer refused to admit that obligation was self-motivated. He tried to negate the potential for internecine war which marks the flip side of self-interested liberty (or fraternity) by placing sovereignty in a monolithic authority which he based upon paternity and property.

Irrespective of the reality of the present government, Filmer argued that "there is and always shall be continued to the end of the world, a natural right of a supreme Father over every multitude."[17] In so arguing, Filmer concluded an infinite and essentially definitive yet nondivine nature to all social and political activity which, like the Tudor idea of order, allowed for a variety of experiences and expressions, but stripped social reality of self-definition. What was so radical about this seemingly conservative theory, however, was that it located ultimate meaning in an authoritarian social and civil will. There is in Filmer's views the harbinger of an idealist, historicist explanation of human activity.[18]

Since the natural duties of a father and a king are "all one," Filmer noted, only a king ruling according to his own will could authoritatively govern society.[19] And it followed, then, that a society could have but one origin, one authority, one will.[20] With Hobbes then, Filmer knew that there could be no shared power. All requisite legislating powers must be arbitrary.[21] The sovereignty alone "and nothing but this, makes a King to be a King."[22] Limited or mixed monarchy, as perpetrated by Pope and "people," negates and, in fact, crucifies real monarchy, he believed.[23] And while he berated mixed monarchy because it was not an ancient doctrine but rather "a mere innovation of policy,"[24] he advocated an equally radical and innovative policy which placed the lawgiver (the king) above the law. Prerogative, he claimed, is above law, and obedience to the king's law supersedes even obligations to divine law. For Filmer, as for Hobbes, "There can be no laws without a supreme power to command or make them."[25]

Even at his most innovative Filmer attempted to substantiate his views constitutionally. There would be no de facto rights for him. Filmer recognized and admitted that the political tensions of his day reflected an outright struggle for sovereignty—a struggle to monopolize arbitrary power; and he sought to evaluate the place of king and Parliament with regard to this struggle constitutionally. He confronted William Prynne and others who denied the king's veto power, his negative voice. The Parliament is certainly not infallible, he claimed, and "laws *chosen* by the Lords and Commons may be unjust."[26] It is the king's duty, then, to distinguish just from unjust law, and for this he must use his "negative voice." Ultimately, however, the veto acted as a constitutional check against the usurpation of legal authority. Some oppose the veto, Filmer acutely recognized, not because it is in itself an innovative, or creative, instrument, but because "it gives him a power to hinder others; though it cannot make him a King, yet it can help him to keep others from being kings."[27]

Filmer radically altered the idea of a social hierarchy based upon natural order. Hierarchy was not natural; rather it signified authoritarian decision making. And without natural hierarchy or order, it followed that there could be no natural liberties or privileges.

> The difference between a Peer and a Commoner, is not by nature but by the grace of the Prince: who creates honours . . . and also annexeth to those honours the powers of having votes in Parliament, as hereditary councellors, furnished, with ampler privileges than the commons: All these graces . . . are so far from being derived from the law of nature, that they are contradictory and destructive of that natural equality and freedom of mankind, which many conceive to be the foundation of the privileges and liberties of the House of Commons: . . . the truth is, the liberties and privileges of both houses have but one, and the selfsame foundation, which is nothing else but the mere and sole grace of kings.[28]

Filmer transcended the traditional idea of order and radicalized the innovative concept of divine right monarchy. He described a social-political order that was as dependent upon individual will as were Hobbes' and Locke's political states. But in Filmer's theory the individual will of the father, not the self-interested individual wills of the brothers, articulated social and political order. And because of this, liberty and government (obligation) still resulted from and reflected the patriarchal nature of society, while with Hobbes, liberty generated political obligation, and government represented, as substitution, social nature.

Selden too recognized the true motives behind the political and constitutional battles in which the king and Parliament engaged. He chastised the king for raising money "illegally"; and he admonished the "Parliamentary Party" for trickery and underhandedness. They claim law when it is on their side, he suggested, and "parliamentary way" when it is against them (pp. 83, 91). Nonetheless, constitutional government depends upon law, he said, and law defines and limits the king's prerogative (p. 112).[29] "The parliament of England has no arbitrary power in point of Judicature, but in point of making lawes" (p. 102).

Constitutional authority resided with the Parliament, not with the king. According to Selden, only Parliament can make law "that was never heard of before" (p. 69). Not only did Selden deny that kings authored or created social order, but he inversely concluded that a king "is a thing men have made for their owne sakes for quietness sake" (p. 61). Kings incorporated no special magic or honor, and they embodied no eternal, finite nature or order. Rather they signified a social need and functioned purposively to serve the individual wills of the citizens of the country: "there is no species of king," Selden argued (p. 62).

Because men contracted government and defined their government by the laws they made, there was no reason to expect that one type of polity would always exist. Selden went a long way toward accepting the existential nature of government. Although he favored custom and common law, he appreciated the mutable nature of political institutions and political ideas. He analytically diagnosed the recent attempts to advance the "King's Supremacy," as a reaction against the claims of Catholic theorists on behalf of the Pope (p. 64). Governments changed from "Commonwealths into Monarchyes and Monarchyes into Commonwealths" (p. 102), he knew; and if men accepted this historical truth, change would be neither as feared nor as cataclysmic as it was.[30] But now, all those who seek innovation, "Presbiters" and others, Selden rightly noted, pretend that they are restoring antiquity (p. 112).

Filmer confronted change as openly as he could. Considering that laws often needed changing because of "some considerable circumstance falling out which at the time of the lawmaking was not thought of," these circumstances are "referred to the aid of the Prince."[31] Filmer accepted both the necessity to change law and the creative power to make law. He too shared in the secular political consciousness which was redefining sovereignty and was objectifying it in the creative act of legislation. Filmer differed from Selden, Parker, and Hobbes in that he settled this creative power upon the king in deference to his natural patriarchal rights. Nonetheless, for Filmer, as for the others, making law and changing law were civil, secular acts authorized by an arbitrary sovereign power.

As Seldon and Filmer made abundantly clear, the developing theory of secular sovereignty, whether it claimed liberty or authority as its source, derived not from natural order but from property.[32] Selden appreciated the self-interested nature of evaluation that held currency during his day. We measure things by their use to us, and others we measure by our conceptions of our self, he claimed (p. 76). Consequently, government must be evaluated by these same standards and contracts of all sorts can have no other foundation than this sense of propriety. "This ye Epitome of all ye contracts in ye world betwixte man & man, betwixte Prince and subject, they keep them as long as they like them and no longer" (p. 37). And from such an appraisal, Selden determined whether men could take up arms against a prince: if we contract, he said, and you (the Prince) get 80 percent and I (the subjects) get 20 percent and you subsequently try to appropriate my 20 percent, I will defend it "for there you and I are equall, both Princes" (p. 136). So whether secular sovereignty represented paternity or liberty (fraternity), as the political theorists of the seventeenth century from Filmer to Locke described it, and as Marx and Freud understood it during the nineteenth century, "it is all a dispute over the inheritance of the paternal estate."[33]

The pamphlet warfare that raged during the early 1640s conspicuously dramatized the "struggle for sovereignty" and the "crisis of the constitution," and underscored, too, the changing political attitudes about order and sovereignty. Royalist and parliamentary sympathizers of varying intensities advanced, rebutted, and redefined positions concerning authority and political obligation in an attempt to articulate a definitive understanding of sovereignty. Little which resembled Tudor commonplace theories about government or social order was posited. Divine right played a minor role also. The controversy upon which much of the theoretical argument centered concerned whether authority resided in law and, hence, in constitutional justification, or in the power to make law and, then, in the efficient political capacity to exercise order and authority. In another way, the struggle for sovereignty was really a last-ditch effort to fix an essential nature and a social basis to secular sovereignty.

Royalists made most use of legal-constitutional arguments in the 1640s; they presented themselves as flexible traditionalists, and they advocated an enlightened, practical constitutional monarchy against what they believed to be blatantly illegal parliamentary aggression.[34] Duddley Digges accused Henry Parker of attempting to alter the constitution for self-interested purposes, and as had Sir Robert Filmer, compared Parker and his ideas with the Jesuit political philosopher, Juan de Mariana.[35] Ferne also associated the Catholics with the Parliamentarians: "For examine your hearts and try if the name of *Parliament* (which is of honorable esteem withall) be not raised to the like excess of credit with you as the name of the *Church* is with the Papists."[36] Ferne and Digges both recognized the revolutionary nature of the contemporary power struggle, and they creatively articulated Royalist positions that transcended traditional and divine right theories of kingship and which in essence defined a constitutional and limited monarchy.

For Digges and Ferne, law limited the king and the subjects. If the king and his opponents reach an impasse, said Digges, the "subject must be obedient to the ordinary law."[37] Ferne went so far as to argue that the king is limited "according to the Lawes" and that the people have the lawful right to deny obedience to unlawful commands.[38] But he claimed, too, that the king historically played the major role in a mixed government.[39] And he denied that Parliament's powers derived from any "first Constitution of this Government," as Herle had suggested.[40] So while the Royalists agreed that king and Parliament shared legislative tasks, they did not agree that they shared power or supremacy.[41] Law limited action but it did not embody sovereign authority. Digges argued that precedent was binding, too, and the attempt by Parker and others to transcend precedent was really an attempt to tie sovereignty to the creative, legislative activity of Parliament.[42] Ferne most inventively refuted parliamentary claims while he limited the monarch's position. He denied that monarchy was the only government possible and he denied that any government was of *jure divino*. But the authorized power to govern was, no matter who exercised it, of God.[43] Government and order then derived not from nature or by correspondence but from power and authority. And since such authority, no matter how limited it might be in reality, was of an original and divine nature, it had to be supreme. So by tying authority to God's original creative "*Dixi*, that silent *Word*, by which the world was at first made," Ferne rejected any parliamentary claims to sovereignty based upon constitutional rights or upon de facto legislative power.[44] And at the same time he attempted a conservative, yet innovative reassessment of government activity, which denied secular sovereignty.

To support this position, Ferne differentiated between monarchy and other forms of government. Like Filmer, he emphasized the original social and paternal nature of monarchy. Indeed, he argued that natural paternal power enlarged becomes monarchy.[45] On the contrary, aristocracy and democracy were of human design only. No contractual agreement could have initiated government, and Parliament's rights must be understood as subsequent to monarchy "and procured by the people for their greater security, not precontrived by them."[46]

Digges established a patriarchal basis for monarchy, too, but taking cues from Hugo Grotius and Selden, he attempted to reconcile the natural, social origin of government with the idea of an original contract. Government began after the original paternal family generated other families. In order to attain their greatest convenience, the divided paternal powers reasonably and voluntarily yielded to the leadership of one common father.[47] Digges did not emphasize the submission involved in this contract because he would not accept the idea that confusion preceded government. The contract, rather than contriving order, reflected nature's sense of things. So while he recognized the contractual nature of government, he maintained that the authority yielded in contracting could never be repossessed.[48] Like Hobbes, Digges stressed the importance of a single source of authority. The one who has authority, he said, protects against divisiveness.[49] There can be no "one state" if the supreme power is divided.[50] Yet unlike Hobbes, he did not see the contracted sovereign as a substitute for the people, representing and fulfilling their native desire. And for this reason, like Ferne, he still understood the "state" as inferior to social nature.[51]

The Royalists confronted the increasing consciousness of self-interest and the relationship between self-interest and public order. Digges presented an interesting opinion which supported the monarch's right to do public good by seeking his private interest alone.[52] He noted that it was useless to strive after one's own private good, however, because to insure order the public good of protection limited private ends. And the monarch's private interest was the price paid for the people's secondary public good.[53] Here, Digges resembled Hobbes more than he did James I, for instance, who had tried to fashion a positive relationship between the king's and the people's good. Although the common order depended on the aggrandizement of the king's private interests in Digges' argument, the two goals were so unrelated that the king appeared alienated from the common good.[54] And in a fashion that suggested agreement with Hobbes' early views in *The Elements of Law*, Digges emphasized the natural proclivity in men to surrender their self-interests, their freedom, and their power in order to insure safety and protection.[55]

Ferne identified order with the public good, "the life of a Commonwealth," and argued that power and authority established this order. And because of the monarch's propinquity to God in "Supreme Power," he would direct his authority toward order.[56] Order, then, rather than reflecting a natural first cause, was a practical end which the king's power authorized. For this reason, Ferne maintained, self-interest, which he believed to be the motivating factor behind resistance to the king, tended toward the dissolution of order and "strikes at . . . Power it selfe," which created order.[57]

It was the awareness of the relationship between self-interest and social utility that formed the foundation for the political arguments which complemented their constitutional-legal positions, and which, in reality, placed Digges and Ferne closer in spirit to their opponents than they might have appreciated. Digges admitted that the government may have perpetrated some recent evils, but he concluded that the English still experienced a better quality of life than any other people.[58] Occasional abuses were a small price to pay for a policy that provided order and general public welfare.[59] Political ends, then, could in some ways justify means, though Digges respected an a priori nature for government. Still he evaluated the existent government on grounds of general satisfaction with what it provided. This was a far cry from a traditional means of evaluation. And it indicated a direction for a total redefinition of social and political purposes along such lines.

More than Ferne or Herle, both of whom he accused of advancing constitutionally illegal arguments about the state being separate from the king or from the people, Philip Hunton recognized human responsibility for political institutions. And he appreciated better than others, too, that change and development characterized such institutions.[60] As did Filmer and the Royalists, he believed that the initial ordinance of government was God's, but he sensed, too, that the "ends of Magistracie" were human. Men constructed political societies. "God by no word binds any people to this or that form till by their own Act [they] bind themselves."[61] By this he comprehended that "Government and Subjection are Relatives" (p. 1). Definition and limitation were not of any divine nature, then, but men, in fact, received what they chose "according to their Relations to the forme they live under" (p. 3). Implied directly in this sense of the nature of government was "a right to alter." Hunton respected no precedents about original and natural constitutions, and he accepted contrived change as well as "natural" development (p. 3). While Filmer and the Royalist theorists discussed obligation in relation to the origin of governmental authority, Hunton was more interested in the relationship between obligation and the source of governmental limitation. A king could limit himself by his own will, Hunton admitted, as James I and Ferne had suggested that monarchs did. But unlike the Royalists, Hunton believed that this was insufficient. There had to be another source of limitation; and this he located in legal restraint. Statutes and common law were composed by the concurrence of the king and the other two estates "so that to be confined to that which is not merely their owne, is to be in a limited condition" (p. 32).

Since he rejected the notion that kings alone made and granted laws, and because he believed that governmental authority resided in the "all-pervading supremacy of legislation,"[62] Hunton adamantly countered the Royalist view about the origin of the king's authority which he maintained came "not from the free determination of his owne will" (p. 12). Where Ferne, for instance, maintained that the king held ultimate power outside of the legislative nature of government and where parliamentary advocates were developing a concept of parliamentary supremacy, Hunton posited, and described, a mixed supreme authority based upon a shared lawmaking responsibility and a contracted set of reciprocal obligations. Carefully, however, he kept the monarch's superiority intact. The contract conveyed a power to the sovereign which could not be reappropriated. "They have divested themselves of all superiority and [have] no Power left for a positive Opposition of the Person of him whom they have invested" (p. 16).[63]

Hunton's *A Treatise of Monarchie* actually described a mixed social order rather than a mixed government as Judson believed. Hunton refused to base government on the social nature of the family as the Royalists had, and he did not grasp the full existential nature of the state as Hobbes did either. Government differed from public society which was "politicall" and from family which was "oeconomical" (p. 1). Government directed political society.[64] The end of government was, of course, the public weal, and a governor had to direct his authority toward this end (p. 4). "The preservation of the Power and Honour of the Governour is an end too . . . but subordinate to the other," the public and social weal, he argued (pp. 4–5). Hunton's government did not represent public society and thereby appropriate public ends into its own as Hobbes' Leviathan did, but it did function as a rudimentarily purposive agent of the public weal and thereby it was effectively separate from public society.

A general difficulty associated with limited monarchies and a particular problem in the early 1640s concerned how to evaluate how satisfactorily the government fulfilled its directive to pursue the public weal. Hunton was imprecise about this. Evaluation was based upon the moral judgments of "reasonable creatures" (p. 18). If something seemed bad to "most," they could ignore it, or petition the government, or resist it. Reciprocal obligations existed but there was no systematically defined manner within government to maintain the positive exercising of the contract. Hunton described a loose secular society with a political government; he did not describe a secular state. In an attempt to appease all sides, he articulated a political society within which king and Parliament shared legislative authority to do the public good.

But Hunton could not avoid attending to the relationship between the king's authority and the people's liberties; and in so doing he began to locate the ultimate power of government in Parliament. As he knew, the "contrivers" of the English "frame of Government" conserved the "Sovereignty of the Prince" as well as the "Peoples freedome" (p. 41). But whereas the monarch cannot contravene this "freedome," the "two Estates" can "moderate and redress the excesses and illegalities of the Royal power . . . by . . . a power to medle in acts of the highest function of Government; a power not depending on his will, but radically their owne" (p. 41).

Not only did Hunton admit the possibilities of an illegal use of royal authority but he indicated, too, an elementary concept of legislative sovereignty. And while the king and Parliament together formed the "Supreme Court" of the country, the Parliament possessed an independent power which could monopolize this legislative sovereignty (p. 42). Now his position and the Royalist views described antithetical appreciations of contemporary political society. Neither, however, realistically appraised the polity, which since Thomas Cromwell's time, at least, had been characterized by a developing delicate balance between social, political processes and legislative, constitutional definition. Each position emphasized one end only of this precariously balanced, embryonic state. But neither position was practical. Only with Hobbes' articulation of a definitive secular sovereign state would a conceptual political philosophy approximate the real polity in which form and process were the same as definition and purpose.

The logical consequence of Hunton's arguments provided a "constitutional" defense of present parliamentary resistance to the king. "The two Estates of Parliament may lawfully by force of Armes resist any persons or number of persons advising or assisting the king in the performance of a command, illegal and destructive to themselves or the publique," claimed Hunton (p. 52). Resistance of this nature, Hunton assured his readers, would restore "Order"; it would not cause disorder, he claimed. Within such a political context, however, the old commonplaces retained only superficial interest and Hunton treated them accordingly. Even the threat of a "Civille War," that traditional bugbear, did not temper his position. If it comes, he argued, it will be the king's responsibility; "yet a temporary evill of war is to be chosen rather than a perpetuall losse of liberty, and subversion of the established frame of a Government" (p. 61). Only a confident appreciation of self-definition could account for this radical disregard for traditional proscriptions. Although Hunton claimed to be restoring liberty and the constitution, it is apparent that for him personal meaning was no longer defined through correspondent social nature. Individuals were responsible for a public order they had created, and the maintenance of it provided them with another opportunity to exercise their creativity rather than to secure their prescribed identities. As he challenged:

> Is it credible that any one will maintaine so abject an esteeme of their authority that it will not extend to resistance of private men, who should endeavour the subversion of the whole frame of Government, on no other Warrent then the King's Will and Pleasure? (pp. 56–57)

Although Hunton presented himself as a Hooker-type compromiser, denying extreme positions and reasoning both sides of the issues, he always ended up advocating radical behavior. He clearly stated that "Religion . . . is not a justifiable cause to take up Armes" (p. 75). Parliamentary force could be justified only "for security of their Privileges, Lawes and frame of Government" (p. 78). Without describing a particular secular sovereign state, then, Hunton still believed that political activity should have political, civil ends only. And in the name of a final and peaceful resolution to the contemporary crisis, he asked the king voluntarily to "suspend the use of his *negative voice*, resolving to give his royall assent to what shall passe by the *major* part of both Houses freely voting, concerning all matters of grievence and difference now depending in the two Houses" (p. 79).

Henry Parker was more straightforward. To save the country Parliament must assume full responsibility for governing, and must "deny the King's negative voice."[65] Charles Herle rebutted Ferne and defended Parliament's taking up arms in defense of king, law, and government "without, and against the *Kings* personall Commands, if he refuse. . . . The finall and casting result of the States judgement concerning what those Lawes, dangers and means of prevention are, resides in the two Houses of Parliament."[66] In denying the king's veto power, and in advocating purposive armed resistance, Parker and Herle acknowledged that private men were responsible for government and for order. Parliament's important role as the institutional representative of this private responsibility could not obfuscate the fact that Herle and Parker described a social and political reality that revolved around private individuals.[67] This signified a major change in the understanding and the conceptualizing of the polity and of politics. "In the thought of these writers, the transition from the medieval to the modern became explicit."[68]

The welfare of the people was the one criterion against which any activity, public or private, could be judged; and in the name of the public good, nearly any activity could be proposed and rationalized. "If Ship money, if the Starre

Chamber, if the High Commission, if the Votes of Bishops and Popish Lords in the Upper House, be inconsistent with the welfare of the Kingdome . . . they [should] be abolisht," declared Parker.[69] He reduced everything to its relationship with the final end of man's benefit. If something positively affected this end, it was good and natural; if it did not, it must be resisted and abolished. Men "erected" the king's dignity to serve and to preserve the "Commonality," he said. The king and the Parliament functioned solely to insure man's welfare.[70]

Such purposive ends redefined origins, then, and expressed a private secular meaning for temporal existence which delivered political order and activity from divine or natural premises and limned them directly as efficient causes. Herle proposed that parliamentary men, in order to protect their own self-interested privileges and properties, defend the nation against the king's self-interested prerogatives. Since men created the king to protect their own welfare, the king's interests, monopolies, patents, and absolute rule, were anathema to the national good.[71] But logically, the individual's privileges and the national welfare described the same parameters.

It was with respect to these considerations that Parker examined the ship money issue. Here the people's liberties and king's prerogatives confronted each other. Parker attempted to abate confusion about the people's privileges and "Freedome." He admitted the king's prerogative, but he pitted it against liberty; "such a prerogative hath been maintained, as destroys all other Law, and is incompatible with popular liberty."[72] Prerogative and liberty are not necessarily irreconcilable, he claimed. But an unequal homology qualified their compatability. The prerogative is "the more subordinate of the two," because the subject's liberty determines the king's strength, he stated. And consequently "the King's prerogative hath only for its ends to maintain the peoples liberty."[73]

In order to moor the ship money case to this pragmatic account of liberty and prerogative, Parker sought a supplementary, but rhetorical, means of evaluation. The people's liberties allowed of no restrictions because "in nature there is more favour due to the liberty of the subject, then to the Prerogative of the King . . . our dispute must be, what prerogative the peoples good and profit will beare, not what liberty the King's absoluteness or prerogatives may admit."[74] Again, then, the ultimate good of the people determined the issue, and any particular position could be advanced or attacked on such grounds. Parker's arguments created an ethics that resembled a personalized version of Bacon's utility and a generalized version of Hobbes' fear of the *summum malum*; and once one established such an ethics, evaluation became a matter of justification and self-interested aggrandizing behavior was easily validated.

Prerogative, then, could only be defended when it profited the people, and when it proved to be "such as the people can not subsist at all without it."[75] Far from being of natural or divine origin, prerogative was nothing more than particular law which gave the king certain privileges not "essential to royalty." And for that reason, at any time, the people who created prerogatives might exercise their same power to annul them.[76]

The people not only created, defined, and limited prerogative, but they assumed a like responsibility for every other political activity and indeed for the polity itself. As authors of society's "Paction," the people "may ordaine what conditions, and prefix what bounds it please," Parker maintained. And this included ending the contract, too, because only those who created it could dissolve it.[77]

Such a life and death responsibility for the commonwealth had to be burdensome, and the nature of all political definition originated with the self-consciousness of this burden. As Parker appreciated, "even that power by which he [king] is made capable of protecting, issues solely from the adherence, consent, and unity of the people."[78] Traditional government insured that the people's welfare was a function of their repressed responsibility. But when self-interest and the self-consciousness of responsibility advanced together, the opportunity existed to construct and to define an "objective" political reality which would assume this personal burden without alienating or redirecting the people's self-interests. Parker believed that Parliament undertook responsibility for social order and public welfare.[79] And the people's self-interest would be accomplished, he argued metonymically, because "Parliament is indeed nothing else but the very people itself artificially congregated; or reduced by an orderly election, and representation, into such a Senate, or proportionable body."[80]

Such an understanding of Parliament's ultimate purpose explains, in some way, why Parker emphasized arbitrary power and unity in government and why he turned to Hooker for help in refuting Calvin on the issue of concurrent jurisdictions in the polity.[81] Government alone ordered society, and to do this efficiently, it required absolute and arbitrary power; "Supream power ought to be intire and undivided, and cannot else be sufficient for the protection of all, it doe not extend over all; without any other equall power to controll or diminish it."[82] It was in this context that Parker began to refer to a "reason of State," and began to locate ultimate sovereignty in Parliament's necessary right to "make use of an arbitrary power according to reason of State, and not confine themselves to meere expedients of Law."[83]

In linking arbitrary power to a "reason of state" and in granting to Parliament the "very people itself artificially congregated," a sovereignty based upon the exercising of such power, Parker came close to articulating a full noneschatological order based upon existential reality alone. Only the people's willful self-repression could redeem and order society. And, indeed, Parker affirmed that "both Houses have an arbitrary power to abridge the freedom of the Subject. . . . To have then an arbitrary power placed in the Peers and Commonors is naturall and expedient at all times."[84] The recognition that sovereignty resided in the power to control law, not in law itself, marked an acceptance of man's creative responsibility for social order, and a consciousness of the self-definitive nature of the historical world. To locate order in the created ordering polity was, indeed, "naturall and expedient" together.

Judson noted that Parker changed from an advocate of law to a defender of "Power."[85] Parliament required total power to make, preserve, and enforce law; and while Parliament was the source of law, power, not law, was the foundation for government and order.[86] Parker and Herle were political theorists, not constitutional thinkers. Where Hunton, for example, still understood a constitutional basis for mixed government, Herle and Parker appreciated only its political benefits which accrued to the people's welfare.[87] The transition from constitutional to political conceptualizations of government characterized a larger shift from an idea of order described by social hierarchy to an idea of order defined by a secular sovereign state. In other words, the change from a medieval to a modern idea of politics and of the polity detailed a switch from the concept of a mixed society determined by law to the articulation of a mixed state which exercised power over law.

Herle's conceptualization of England's government as a mixed monarchy indicates this transition. "The Monarchy, or highest power is itself compounded of 3 Coordinate Estates, a *King*, and two *Houses* of Parliament," he said.[88] This signified an alteration of the political relationship that had shaped the traditional and mixed society. Where the king *in* Parliament had traditionally constituted the whole Parliament, now the king *and* Parliament instituted the monarchy. The Parliament, then, was in no sense subordinate to the king, because the Commons, the Lords, and the king were each a "Part of the Monarchy," or the state. And consequently, Parliament is a "*coordinate part* in the Monarchy, or highest principle of power, in as much as they beare a consenting share in the *highest office* of it, the *making* of *Lawes*."[89]

Society and state had become quite distinct. No longer did social nature and social order prescribe the frame of government. As Herle made clear, men who were social subjects of the king were his equals in Parliament, in the state.[90] Parker, too, abstained from exercising the old correspondence theories. The fact that women served men and that children served their fathers had no bearing on political relations. In the state, between man and man, he claimed, all relations defined reciprocal and equal obligations whose end was the common good.[91] Social nature had lost its monopoly as the determining force of human identity and of self-society relations, and the state had begun to define a locus of relationships that ultimately transcended traditional ones in social and psychological importance. As Parker calculated, a subject's treason against his prince was far less offensive than a princes' oppression of a subject.[92] The first reneged only on a social obligation, while the second disowned a human and political contract.

But when political push came to ethical shove, Parker and Herle retreated behind the traditional banners of the original constitution, of the "beames of humane-reason," and of the fundamental inclinations of nature and of natural order. Parliament's arbitrary power, even though it directed government toward self-preservation and order, derived ultimately from nature, Parker believed. And order, the end of government, emanated from God and was of a "sublime and celestiall extraction."[93] And while he approached a Hobbesian appreciation of secular sovereignty based upon existential order, Parker still differentiated between God's natural essential order and man's human jurisdiction. Absolute power, then, Parliament's means to an ordered and secure commonwealth, linked God and history. As Parker knew, according to "internal operations of the Deitie we ought . . . to ascribe prioritie of Order to infinite power."[94]

Herle was even more cautious, and even less capable of evaluating change and order-in-themselves, than Parker. An original constitution "framed" the mixed monarchy, he claimed. This foundation, "the fundamental law" of the nation, was "in every severall raigne confirm'd both by mutuall Oathes between King and people, and constant custome."[95] In this fashion Herle resisted bearing the human burden for social and historical change, and while he advocated parliamentary sovereignty, he denied its self-definitive nature. By "*Reason*" and by "*Wisdome* of

State" Parliament voted to reiterate the original contrivance, he argued.[96] Parliament's laws, then, interpreted the ancient constitution and energized its potential for creating order. And consequently, the power of the "mixed monarchy" to determine public safety derived from an original contract. Parliament's immediate and sole responsibility for government and for order resulted not from an usurpation of power or a redefinition of function, but rather from an abrogation of obligation by the king. Because the king abandoned his duty to the "*Reason* of the Kingdom," the Parliament was not similarly exempted from its duty. And since "there is more need of declaring old Laws then of making New," Parliament would no doubt save the day.[97]

In the *Leviathan*, Hobbes modified his political psychology. In his earlier works he derived political nature from the nature of individuals. In *Leviathan* he began to understand that political nature reflected "the formal structure of the relations between individuals."[98] As each individual moves to preserve himself in a world full of like-motivated individuals, he creates natural laws which define a self-other nature for social reality, and which prescribe means to ends. Such laws compelled order because certain psychological principles defined them. Natural law reflected the psychology of man in motion and of motive men in conflict. As J. W. N. Watkins concluded, Hobbes reduced the gap between desires and prescription and reduced political nature to a single psychological system of "basic wants and proven hypothetical imperatives."[99]

Reason and experience direct the passionate motion for self-preservation toward peace and order, Hobbes argued. The human laws of nature restrict man's natural rights and simultaneously accomplish the ends of motion by defining order. Hobbes' laws of nature, then, were functional and pragmatic rather than moral.

Man's rights of nature direct him "to use his own power, as he will himself, for the preservation of his own nature."[100] And "every man has a right to everything; even to one another's body" (p. 85). Liberty, then, manifests freedom from external impediments to motion, power, and will. By right, Hobbes argued, man could exercise this liberty and search after preservation or he could obligate himself to "forbear" his *own* search (p. 84). And herein conflict continued. But reason and the law of nature bound man to one motion alone, "so that law, and right, differ as much as obligation, and liberty; which in one and the same matter are inconsistent" (p. 85). Hobbes, then, recognized that liberty and obligation were two means to the same end, but that since liberties naturally conflict, obligation was a surer and a safer path to take. And by definition of the law of nature, the one path ruled out the other. Since order and preservation required an understanding of the natural relations between man and man, Hobbes argued that a man must

> be willing, when others are so too, as far-forth, as for peace, and defense of himself he shall think it necessary, to lay down this right to all things: and be contented with so much liberty against other men, as he would allow other men against himself. . . . To lay down a man's right to anything, is to divest himself of the liberty, of hindering another of the benefit of his own right to the same. (p. 85)

Man could either "renounce" his rights or he could "transfer" them. But only in the latter instance did he oblige himself to the body to whom these rights would now belong. And of course such an obligation denied man the liberty to hinder this "other" body of rights. When all men together transferred their rights and thus obliged themselves they manifested this voluntary action in a sign—a political legislative sovereign state which represented, in its substitute body, these transferred rights. The supremely sovereign act, then, creatively negated liberty, or self, and redirected its own generative impulse toward order. Hobbes redefined man's relationship to order by deriving sovereignty and obligatory law from the sublimated rather than from the repressed nature of man.[101]

The act of transferring these rights of nature, argued Hobbes, was a "contractual" one. "To convenant, is an act of the will; that is to say, an act, and the last act of deliberation" (p. 90); the appetite to order becomes contractually manifest in the sovereign's will alone—in its appetite. The creation of a civil sovereign power was simultaneously, then, the beginning of the execution of contractual obligations (p. 94). And a true commonwealth instituted a political reality whereby all power is conferred upon one man or an assembly of men "that may reduce all their wills, by plurality of voices, unto one will . . ." (p. 112). Each member of the commonwealth transferred his authority to the sovereign on condition that each other member did so too, and each realized that he was in fact the original author of—was responsible for—the sovereign's actions. The Leviathan, or "mortal god," as Hobbes called it, did not will for the people, but shaped and directed their generating wills through its own creative will. And because of this, self-interest was preserved without contravening public utility. Indeed, "in a commonwealth . . . not the appetite of private men, but the law, which is the appetitie of the state, is the measure" of good and evil (p. 446). The one appetite then provided for private and public ends. Hobbes refuted all the recent arguments that divorced the people's liberties from the king's prerogatives. "The good of the sovereign and the people, can not be separated," he said (p. 277). By focusing upon the psychological nature of motivated self-other relations, Hobbes redefined a public-private nexus for order and identity. Now, however, man authored this mutually purposive endeavor. "There could be no question of imposing a system of values from outside or from above."[102]

As was the case in his conclusions about liberty and obligation, about rights and laws and about private desires and public interest, so too did Hobbes maintain that social nature could be either fraternal or patriarchal, but not both at the same time (pp. 101–102). There could be no conception that "two, or more things can be in one, and the same place at once" (p. 17). Hobbes located political order discursively in definitional order. He had no use for the burdens inherent in Shakespeare's oxymoronic, complementary vision. He did not reject alternatives, however, but founded order upon the individual's responsible choosing of one alternative, without mandating that the same choice be made over and over again. He rejected the traditional and the radical Christian perceptions of one order only. And he denied the Shakespearean vision of complementarity. In their place, Hobbes substituted a new perceived order by deifying the self-reflective and self-consciously definitive one choice.

CONCLUSION

The Self-Consciousness of Consciousness: The Perceived Order of Modernity

> Truth is not found in certain historical agents nor in the achievement of theoretical consciousness, but in the confrontation of the two, in their practice, and in their common life.
>
> Maurice Merleau-Ponty,
> "Materials for a Theory of History"

> There is a marriage (in heaven) between psychoanalysis and the mystical tradition; combining to make us conscious of our unconscious participation in the creation of the phenomenal world.
>
> Norman O. Brown, *Love's Body*

Intellectual history can be definitionally reduced to any of a variety of phenomena or contexts. This has not been the intention in this book. In this brief conclusion, I would like to reiterate the relationship between consciousness, order, and social reality which this book has assumed, to suggest some considerations to widen the possibilities of the interpretation, and to look once again at Hobbes and representative sovereignty in relation to the intellectual history of modern Western social reality.

As I stated in the Introduction, consciousness relates to social reality by limiting possibilities—this is order. It operates in social reality as "a metaphor of our actual behavior":[1] behavior that is culturally and historically determined (constructed). An investigation of the relationship between consciousness, order, and social reality requires that we study consciousness in history—or the history of consciousness—and may be formulated along the lines of Kuhn's paradigms. Consciousness, then, appears as the quality of perceiving order in things. And the history of consciousness is concomitantly cultural history—a history of modes of perception and modes of communication; it is thus a history of different "orders." Consciousness reemphasizes all phenomenal experience through language and thereby constitutes a perceptible, ordered reality for this experience. And

> order is, at one and the same time, that which is given in things as their inner law, the hidden network that determines the way they confront one another, and also

> that which has no existence except in the grid created by a glance, an examination, a language: and it is only in the blank spaces in this grid that order manifests itself as already there.[2]

A change in the idea of order is a change in consciousness. The self-defined, articulated, representative order contextualized in the mid-seventeenth century signified the development of self-consciousness. Order now resided in definition, in the conscious naming of things which reduced objective reality to a series of identities. Order was no longer a perception of consciousness; it was now a responsible articulation of the self-consciousness of consciousness. And as consciousness had begun to perceive order in social reality when language developed as an abstract tool, now the self-consciousness of consciousness began to order and define a modern world as language (words) gained "literal" monopoly rights for the representation of reality. Again, the history of consciousness reveals that consciousness reflects *history as order* and that the self-consciousness of consciousness represents *order as history*.

Such as emphasis on identity, definition, names, on the "literal," characterized the intellectual history and the historical psychology of the modern Western world, or the seventeenth, eighteenth, and nineteenth centuries. In Kuhnian terms, the recognition by consciousness "of the essential *differentnesses* among all particular things . . . led to an abandonment of that mode of discourse founded on the paradigm of resemblance," which defined the traditional idea of order.[3] "In place of sympathy, emulation, agreement, and so on, the seventeenth century opted for the categories of order and measurement, conceived in essentially *spatial* terms."[4] Discrimination described the mental activity that began to appreciate the temporal nature of existence during the early seventeenth century. And self-definition began to contrive a public order which secured private identity within the temporal space of existential reality.

Within this context it is clear that an intellectual history of order in Renaissance England, of necessity, dealt with the development of self-consciousness and self-definition and with the relationship between private and public during these years. Kuhn's general paradigm model allowed for an examination of this history within the configuration of the larger revolution in consciousness and order. "General revolutions in sensibility . . . are results of an overloading of a received mechanism of encodation beyond its capacity to function at all."[5] And as historical psychology as well as modified Freudian analysis reveals, "anxiety can be experienced when disrupted social conditions threaten the adequacy of ego's efforts to integrate cognitive, perceptual and other functions with changing reality, making action difficult and ineffective."[6]

A breakdown in the social world or in the perceived order that structures the objective world and facilitates positive self-other relationships necessitates an anxiety-informed attempt to redefine self and society. Shakespeare's vision problematizes what theoretical analysis announces: "Men will fight either to preserve the traditional symbolic codes that have regulated their lives or to establish new codes . . . men cannot act, and will not live, in the absence of such symbolic codes."[7]

"The guiding principle in redemptive history is the election of a minority for the redemption of the whole. Otherwise expressed, it is the principle of representation."[8] By relocating order in a representative sovereign state Hobbes redeemed self-in-society noneschatologically. It was the state as it manifested the creating principle of election, not the state-as-being, which ordered reality and provided a nexus for positive self-society relations. For Hobbes, order resided in definition, and definition was a function of continual choosing and creating. Meaning, order, and redemption for self-in-society were *becoming*, not being.

As Voegelin said: "Articulation . . . is the condition of representation. In order to come into existence, a society must articulate itself by producing a representative that will act for it."[9] Articulated representation, Hobbes' secular sovereign state, is sexual, is creative. The key to secular sovereignty rests in its being both created and creative. Representation as substitution. Man creates (articulates) a self-defined order (state) and simultaneously transfers his creating potential, his self-definitive capacities, to his creation which uses them for the benefit of self-in-society (legislative sovereignty). The articulated Leviathan manifests the self-definitive, utilitarian, will not to will. Man's gnostic, creative, self-transcendent possibilities remain alive within the sovereign representative state, but now they appear to be functioning positively rather than exacerbating the isolating tendencies of self-definition.

Michael Oakeshott describes three idioms of morality which he maintains follow each other in European history: the "morality of communal ties," the "morality of individuality," and the "morality of the common good."[10] The first describes a period when human beings are solely members of society and in which all activity is communal activity. "Here, separate individuals, capable of making choices for themselves and inclined to do so, are unknown." Participation in social activity alone defines good conduct. "It is as if all the choices had already been made and what ought to be done appears, not in general rules of conduct, but in a detailed ritual from which divergence is so difficult that there seems to be no visible alternative to it."

The second idiom characterizes that moment in which human beings are recognized as and have come to recognize themselves as separate and sovereign individuals. "It is the morality of self and other selves." Individual choice is preeminent. Relationships determine action and "morality is the art of mutual accommodation."

Finally, in the third stage, "human beings are recognized as independent centres of activity," but behavior that represses self-interest in favor of social benefit defines the locus of ethical approbation. Social good is the sole common enterprise and "morality is the art in which this condition is achieved and maintained."

According to Oakeshott, Hobbes explored the confines of the second idiom only. But if it is to be allowed that *Leviathan* discursively confirmed the "morality of individuality," it must also be pointed out that it also critiqued that idiom self-consciously, and as it were indicated a redefined "social" morality, the "morality of the common good." But modern morality is much more artifice than "art," and its history is pretentious. Perhaps only now can we self-consciously and ironically reflect upon that magical, cultural product, which the great conjurers Shakespeare, Bacon, and Hobbes textually created to be the responsible authority for social reality—individual man. But like all great magicians they perfected that most popular trick—no sooner had they created man than they made him disappear. We are perhaps just now able, reflexively, to acknowledge and to applaud their act.

> As the archaeology of our thought easily shows, man is an invention of recent date. And one perhaps nearing its end.[11]

The *logos* of modern Western social reality, then, is not spiritual, but historical: order as history. Man self-consciously articulates secular order and chooses to translate responsibility from self to state in representation. At the same time that he transfers this responsibility, he alienates self from representative. Yet by the nature of its articulating capacity, self-consciousness retains original authorship, and consequently, authority for social order. Man, then, redefines self just as he redefines God. Both man and God become original creative authors who have transferred their responsibilities for order to representatives. From the divine cosmos to man to the sovereign state—and then, from Hobbes to Freud, the history of modernity is the history of the self-consciousness of consciousness; and it is definitively (deceptively) psycho-logical.

> Vicarious satisfaction: the deed is both theirs and not theirs. On this self-contradiction, this hypocricy, this illusion, representative institutions are based.[12]

And modern Western history, too.

NOTES

Preface

1. E. M. W. Tillyard, *The Elizabethan World Picture* (New York: Vintage Books).

2. Dominick LaCapra, *Rethinking Intellectual History: Texts, Contexts, Language* (Ithaca, N.Y.: Cornell University Press, 1983), p. 16.

3. A conspicuous example is Franco Moretti, "'A Huge Eclipse': Tragic Form and the Deconsecration of Sovereignty," *Genre* 15 (1982): pp. 7–40. Using a functionalist analysis, Moretti argues that the task performed by "tragic form" (in the Elizabethan and Jacobean periods) was the "destruction of the fundamental paradigm of the dominant culture" (p. 7). While this might be a useful point to begin a more refined argument, it is used here instead to generate logically and historically fallacious conclusions. "Tragedy disentitled the absolute monarch to all ethical and rational legitimation . . . it made it possible to decapitate him" (pp. 7–8). Instead of examining the marginal in reference to the taken for granted, Moretti applies a reverse reductionist analysis. "At bottom, English tragedy is nothing less than the negation and dismantling of the Elizabethan World Picture" (p. 12). And further, he denies the same self-conscious critical experience and function to other discursive phenomena. For instance, while he casually identifies Hooker as "the great codifier of Elizabethan ideology" (p. 11), he argues that only "tragedy is truly modern, truly rigorous" (p. 14). Deconstruction must have its limits after all.

4. LaCapra, *Rethinking*, p. 68.

5. In a similar vein, I believe, J. G. A. Pocock in "Languages and Their Implications: The Transformation of the Study of Political Thought," in *Politics, Language and Time: Essays on Political Thought and History* (New York: Atheneum, 1971), advises that we must seek to make the "implicit explicit and to find levels of meaning in a man's thought which he did not directly express and of which he was not consciously aware." We must carefully "indicate the historical moment at which the implicit is seen as becoming explicit" (p. 33).

Introduction

1. Astraea is, of course, Elizabeth and is the symbol for imperialist mysticism surrounding the queen: the "Virgin of the Golden Age." See Frances Yates, "Queen Elizabeth as Astraea," *Journal of the Warburg and Courtauld Institutes* 10 (1947) pp. 27–82, and see also Frances Yates, *Giordano Bruno and the Hermetic Tradition* (London: Routledge and Kegan Paul, 1964), pp. 289, 392.

2. Michael Walzer discusses the sixteenth-century use of organic analogy as a means to sustain the life of the body politic. This was an end in itself. Although there was security about personal salvation, there was no idea of political salvation. Order, as a means of social redemption, was necessary because of the prevalence of a cyclical

view of history. Michael Walzer, *The Revolution of the Saints: A Study of the Origins of Radical Politics* (New York: Atheneum, 1971), p. 178.

3. Thomas Norton and Thomas Sackville, *Gorboduc*, in *Medieval and Tudor Drama*, ed. John Gassner (New York: Bantam Books, 1963), p. 453.

4. Christopher Marlowe, *Doctor Faustus*, in *The Complete Plays*, ed. Irving Ribner (New York: Odyssey Press, 1963), I. i. 23–26, 45–47, 49.

5. Eric Voegelin, *The New Science of Politics* (Chicago: University of Chicago Press, 1952), p. 180.

6. Hayden White, *Tropics of Discourse: Essays in Cultural Criticism* (Baltimore: Johns Hopkins University Press, 1978), p. 5.

7. Much of the argument in these last three paragraphs, and again in the Conclusion of this book, comes from, and is developed more fully in, Stephen L. Collins and Alice L. Savage, "Consciousness, Order and Social Reality: Towards an 'Historical' Integration of the Human Sciences," in *Humboldt Journal of Social Relations* 11:1 (1983–1984), especially pp. 90–101.

8. For an argument in favor of the medieval origins of modernity, see Voegelin, *New Science*. A most persuasive defense of the eighteenth-century origins of modernity can be found in Peter Gay, *The Enlightenment: An Interpretation*, 2 vols. (New York: Alfred A. Knopf, 1966–1969).

9. In using the designation "gnostic," I have in mind the usage of Voegelin, who describes the gnostic experience as a means to help man secure his hold on faith and the meaning of existence by "an expansion of the soul to the point where God the final cause is drawn into the existence of man." Following this, Voegelin explains that secularization is the consciousness of man as god. *New Science*, pp. 124–125.

10. In *Radical Tragedy: Religion, Ideology and Power in the Drama of Shakespeare and His Contemporaries* (Chicago: University of Chicago Press, 1984), Jonathan Dollimore argues that it is, in fact, in the Enlightenment that a form of "abstract individualism" emerges which distinguishes "the individual from society," and gives priority to the individual. "The individual becomes the origin and focus of meaning" (p. 250). He notes that traditionally the individual was constructed "as a binary function of the universal" (p. 159) and that when (in the Renaissance period) the cosmos was destabilized, so was the individual. He mistakenly, I believe, emphasizes heliocentricity as "the first great decentring of man" (p. 160). Rather, it was the subsequent geometrization of space—acentricity—that decentred man. Nonetheless, he is correct to point out that it is consequently wrong to imagine that with the "deconstruction of metaphysics (God) the particular (Individual) was foregrounded in all its intrinsic uniqueness" (p. 159), but his focus on "the emergence in the Renaissance of a conception of subjectivity legitimately identified in terms of a materialist perspective" (p. 249) is unsupportable. He claims that only a materialist (Marxist) perspective "invites a positive and explicit engagement with the historical, social and political realities of which both literature and criticism are inextricably a part" (pp. 249–250). He links humanism, essentialism, idealism, existentialism—indeed, he "dumps" all nonmaterialist (Marxist) interpretations together—with advocating "the idea that 'man' possesses some given, unalterable essence which is what makes 'him' human, which is the source and *essential* determinant of 'his' culture and its priority over conditions of existence" (p. 250). But not every nonmaterialist interpretation posits an a priori essence to 'man' which makes him what he is absolutely, and not every non-Marxist is a Husserl. And while the individual does not immediately create a world that replaces the "destabilized" cosmos, he does struggle to do so and comes to terms with instability in a way that "meaningfully" redefines self and society. By concluding that Marx "mandates a rejection of 'essence,' 'nature' and 'man' as con-

cepts implicated irredeemably in the metaphysic of determining origin" (p. 251), Dollimore retreats from the "historical" interpretation he claims to monopolize. 'Man' does have a 'human' history, however brief, and it is the beginnings of this history (which terminates at the end of the nineteenth century—the history of "modernity") that I attempt to illuminate. Indeed, it is post-modern identity (and history, too) that is difficult to understand as 'man' made.

11. Walzer, *Revolution*, p. 18.

12. Walzer's views certainly have not gone unchallenged over the years. His emphasis on Puritan ideas and attitudes—his supposed idealism—and his position that ministers led the movement well into the seventeenth century have caused the most discussion. The best alternative interpretation remains Patrick Collinson, *The Elizabethan Puritan Movement* (London: Jonathan Cape, 1967). For an interesting, more recent, comparative study in historical sociology, see Mary Fulbrook, *Piety and Politics: Religion and the Rise of Absolutism in England, Würtemberg and Prussia* (Cambridge: Cambridge University Press, 1983). I believe that Walzer's general assumptions about Puritanism remain the most enlightening for an inquiry into ideology and the idea of order.

13. J. G. A. Pocock questions Walzer on this very point. Pocock suggests that there is an aspect of the Puritan personality that is "deeply involved in the secular order." I think Pocock is correct, and I think this is understood best when one keeps in mind Voegelin's ideas about gnosis and secularization (see note 9). See J. G. A. Pocock's chapter on England in *National Consciousness, History and Political Culture in Early-Modern Europe*, ed. Orest Ranum (Baltimore: Johns Hopkins University Press, 1975), p. 107. In another vein, Lawrence Stone has identified two competing "world views" during this period: the "Puritanically ascetic and the secularly sensual." After 1660, he maintains, the former collapsed. See *The Family, Sex and Marriage in England 1500–1800* (New York: Harper & Row, 1977).

14. People chose different means to escape the anxiety that resulted from the dissolution of the traditional concept of order. Those oriented toward religion or philosophy attempted to define security around a revitalized or a fresh transcendental truth. Political and commercial people began, on the other hand, to find purpose and security in a conscious acceptance of the worth of their temporal vocational existence. In most situations, however, the deeper psychological forces which imbued the restless creative challenges in all fields appear to be similar. Walzer, too, recognizes that some people became saints "if their social and personal experiences had been of a certain sort . . . Puritanism . . . is one possible response to the experiences of disorder and anxiety . . . it is one possible way of perceiving and responding to a set of experiences that other men than saints might have viewed in other terms." Walzer, *Revolution*, pp. 308–309.

15. Again Walzer's book illuminates here. Puritanism was dependent upon anxiety. Its existence suggested an "acute fear of disorder . . . a fear attendant upon the transformation of the old political and social order." Walzer argues further that Puritanism lost force when anxiety lost sway. With this in mind I believe that there were other, more dynamic, self-creative forces than Puritanism that helped shape the modern, utilitarian, secular state, and consequently created a social-political expression of a self-defined optimistic order. Puritanism, then, is a less-than-defining alternative because it is an example phenomenon rather than the structurally dynamic force that energized all the alternatives. Walzer, *Revolution*, p. 303. See also Zevedei Barbu, *Problems of Historical Psychology* (London: Routledge and Kegan Paul, 1960), p. 183, for a discussion of Puritanism as a "reaction formation."

16. See the article by Pocock for a comment about the necessity to reassess Elizabethan-Jacobean political thought along these lines, in *National Consciousness*, p. 104.

17. For a fuller elaboration of the meaning and the implications for self and society of this traditional concept of correspondence, see Chapter 1. For the clearest recent examination of this subject see Lacey Baldwin Smith, *Treason in Tudor England: Politics and Paranoia* (Princeton: Princeton University Press, 1986), ch. 4.

18. Voegelin, *New Science*, p. 179. For an interesting interpretation which argues that for Hobbes, history, diachronic experience, *is* eschatological in derivation, see J. G. A. Pocock, "Time, History and Eschatology in the Thought of Thomas Hobbes" in *Politics, Language and Time.*

19. J. G. A. Pocock, "Civic Humanism and its Role in Anglo-American Thought" in *Politics, Language and Time*, p. 81.

20. Eric Voegelin defines two terms which seem to me to aid an understanding of what is being discussed here: *microcosmos*—"the symbolization of society and its order as an analogue of the cosmos and its order"; and *macroanthropas*—"the symbolization by analogy with the order of human existence that is well attuned to being." *Order and History, Vol. I: Israel and Revelation* (Baton Rouge: Louisiana State University Press, 1956), p. 5.

21. For a full discussion of the Tudor idea of the commonwealth, see W. R. D. Jones, *The Tudor Commonwealth, 1529–1559* (London: University of London, Athlone Press, 1970). In his conclusion, Jones discusses the changing conceptions of state into the seventeenth century. His comments are enlightening.

22. Voegelin defines transcendental and existential reality in *New Science*, p. 54.

23. W. H. Greenleaf considers some of these points in *Order, Empiricism and Politics* (London: Oxford University Press, 1965), p. 10. Greenleaf, however, suggests that the divine right Royalists of the Stuart years (his "Order theorists") are expressing a similar political vision to that of the Tudor order theorists. He argues, then, that the conflict between "order theorists" and "empiricists" embodies the spirit of the first half of the seventeenth century. He reduces the innovative aspects of the Royalist arguments, implies a static "order theory," and generally conflates various approaches to political and social order. I hope to show that early seventeenth-century political theory, whether Royalist or not, was basically innovative, and that if there is any "spirit of the age," it is to be found in the underlying motives that were a fillip for all the political theorizing.

24. Isaac Kramnick, "Reflections on Revolution," in *History and Theory*, 11:1 (1972): pp. 26–63.

25. Ibid., p. 33.

26. Pocock, "Languages and Their Implications," p. 13.

27. Kramnick, "Reflections," p. 34.

28. Pocock, "Languages and Their Implications," p. 13.

29. Hayden White, "Foucault Decoded: Notes from Underground," *History and Theory* 12:1 (1973): p. 34. Kuhn's theory suggests that we can treat the history of ideology as a political and a linguistic process. Pocock identifies the value for the historian of "defining the 'paradigm' both in terms of the intellectual (heuristic) function it performs and in terms of the authority, both intellectual and political, which it distributes as between human actors in a social system, we acquire two sets of criteria by which to define the language in general and the paradigm in particular, in which we are interested, in their social context and their historical concreteness." See "Languages and Their Implications," p. 14. He goes on to note that he diverges from Kuhn by treating paradigms in strictly "linguistic terms." He views the verbal para-

digm as "a historical event or phenomenon to which there can be many responses" (p. 15). It appears to me that in a general way such a methodology is useful in interpreting the transformation in the idea of order that we are involved in.

30. Thomas S. Kuhn, *The Structure of Scientific Revolutions* (Chicago: University of Chicago Press, 1970), p. 86.

31. Ibid., pp. 86–87.

32. Ibid., p. 77

33. J. A. W. Gunn has written a detailed account of the relationship between private and public interests in political and economic life and thought after 1640. Gunn realizes that the conflation, to as great an extent as possible, of these apparently opposing interests, was the key to the successful and orderly English political and social life that followed upon the transitional civil war years. He also suggests that "the growing accommodation between particular interests and those of the public . . . were to a degree dependent on new perspectives in human psychology." He does not appear interested in how or why these necessary new psychological perspectives developed. It is clear that these new perspectives were fashioned early in the seventeenth century when the articulate Englishman began to solve the private-public controversy in terms of a newly developing personal identity. J. A. W. Gunn, *Politics and the Public Interest in the Seventeenth Century* (London: Routledge and Kegan Paul, 1969), p. 55.

Again, Kuhn's concept of paradigms is useful here and again, Pocock clarifies its use. Paradigms help "to provide the scaffolding for the assertion of the individual ego's identity as a political being." Indeed, "a conventional set of paradigms cease to satisfy by their performance of this function, and a conceptual revolution occurs so that new paradigms assert new modes of identity in new ways." See "On the Non-Revolutionary Character of Paradigms: A Self-Criticism and Afterpiece," in *Politics, Language and Time*, p. 277.

34. Robin M. Williams, Jr., "Change and Stability in Values and Value Systems," in *Stability and Social Change*, eds. Bernard Barber and Alex Inkeles (Boston: Little, Brown, 1971), pp. 124–125.

35. Two simple examples will illustrate this. Huizinga's discussion of fourteenth- and fifteenth-century France and Burgundy limns how chivalry had lost its original vitality by that time. See Johan Huizinga, *The Waning of the Middle Ages* (Garden City, N.Y.: Doubleday Anchor Books, 1954). A look at the history of usury in Europe shows that throughout the Middle Ages all types of interest were anathematized. By the early seventeenth century in England views were changing. Soon merchants and other businessmen were advocating interest as a valuable means to finance business and thus compete with foreign trade. See Chapter 4.

36. Arthur O. Lovejoy, *Reflections on Human Nature* (Baltimore: Johns Hopkins University Press, 1969), p. 94.

37. Ibid., p. 94. The same can be applied to the Elizabethan public's popular reception of Marlowe's protagonists. For a discussion of Shakespeare and Marlowe, see Chapter 2.

38. For a full discussion of historical psychology, see Barbu, *Historical Psychology*. Barbu concentrates on sixteenth- and seventeenth-century England to show that "the mind is an integral part of the historical process. This means that it is determined by, and at the same time, determines this process" (p. 14). See also Lacey Baldwin Smith, *Treason in Tudor England*, ch. 3, for a sensible introduction to the theme of identity in sixteenth-century England. The most influential recent study of the "self" in this period is, of course, Stephen Greenblatt, *Renaissance Self-Fashioning: From More to Shakespeare* (Chicago: University of Chicago Press, 1980). Greenblatt crea-

tively situates "selves" in reference to ideologically transmitted forms of power, institutionally particularized. He tends to reduce "self" to role(s) and "fashioning" to playing. He begins and ends his essay by stressing the pressure that objective reality exerts on individuals. He maintains that "there is considerable empirical evidence that there may well have been less *autonomy* in the sixteenth century than before, that family, state and religious institutions impose a more rigid and far-reaching discipline upon their middle-class and aristocratic subjects" (p. 1). The "self-fashioning" he examines intersects with, and becomes complicated by, the institutionalization and rigidification of social expectations during the century. But self-fashioning is not the only context for identity. By emphasizing role-playing as a valid psychological response to unsettling social change, Greenblatt privileges an ideologically dominated self, and brackets the view that decorum, role-playing, institutional demands are responses to confirming *and* transforming self-conscious evaluations as much as they are parameters for subjective expression. And certainly, from the late sixteenth century on, the fashioned self indicates self-conscious definition as much as it does modeling.

39. Barbu, *Historical Psychology*, p. 158.

40. Ibid., p. 181.

41. The strong Renaissance magical influences on *The Tempest* and on other late plays is now being fully considered. Frank Kermode's introduction to the Arden Shakespeare edition of *The Tempest* (London: Methuen, 1964), is a fine, penetrating discussion of this subject. Of greatest importance is Frances Yates' *Shakespeare's Last Plays: A New Approach* (London: Routledge and Kegan Paul, 1975). Yates follows up on her earlier work by suggesting close affinities between Hermeticism, Renaissance magic, Rosicrucianism, and, of course, Shakespeare. As I will show throughout, sympathy for Hermetic or magical concepts was more widespread than is generally thought.

42. In his *New Science*, Eric Voegelin is concerned with the consequences of, as he puts it, "a fall from faith." What happens when a large number of people in a society no longer are secured by the given religion, ideology, or social truth that ordered that society? These "agnostics," as Voegelin identifies them, experience a profound psychological fall. Voegelin's comments here are enlightening for a fuller appreciation of the deep psychological content of Shakespeare's work, and also of the possible alternative faiths and beliefs (such as Puritanism or intellectual Hermeticism) that offer a new promise of redemption.

> A man can not fall back on himself in an absolute sense, because if he tried, he would find very soon that he has fallen into the abyss of his despair and nothingness. . . .
>
> The fall could be caught only by experiential alternatives sufficiently close to the experience of faith that only a discerning eye would see the difference, but receeding far enough from it to remedy the uncertainty of faith in the strict sense. (p. 123)

The usefulness of the paradigm model of interpretation is evident here again and Pocock's remark suggests a basic theme which I attempt to elaborate: "To rebel against existing paradigms is indeed to go in search of new ones; but it is also to assert what it is like to be without them, to experience the terror and freedom of existential creativity." Pocock, "On the Non-Revolutionary Character," p. 278.

43. Two groups which can be cited as working, each in its own way, to define a new social order that reflected its own individual values were the Parliament men and merchants of Elizabethan and Jacobean England. G. R. Elton, in *The Tudor Constitution* (Cambridge: Cambridge University Press, 1968), reminds us that "it thus appears that opposition in Parliament . . . grew into something regular, organized and persis-

tently troublesome in the reign of Elizabeth. The reasons may be found in part in the developing self-consciousness of the Commons and the presence there of a Puritan group, in part in the fact that from 1558 the Crown ceased to lead the country into new ways but tried to prevent change" (p. 303). Misselden, Malynes, Mun et al., in pamphlet after pamphlet, propagandized the value and worth of the merchant to a society which still liked, verbally at least, to imprecate merchants. The merchant's success in convincing society that private gain would also be public gain did much to make secular values palatable. For a fuller discussion of merchant self-consciousness, see Chapter 4.

44. Yates, *Giordano Bruno*, p. 156.

45. Ibid., p. 178. Also see Frances Yates, *The Rosicrucian Enlightenment* (London: Routledge and Kegan Paul, 1972), for a full treatment of the interrelatedness of the Hermetic and the political worlds in early seventeenth-century Europe.

46. Yates, *Giordano Bruno*, p. 151.

47. See Herbert Butterfield, *The Origins of Modern Science* (New York: Macmillan, 1957).

48. No better account of this subject has been written than Keith Thomas' *Religion and the Decline of Magic* (New York: Charles Scribner's Sons, 1971).

49. Another experience that lends itself to this type of psychological interpretation is that of the English "virtuosi." According to Walter E. Hilton, Jr., "The English Virtuoso in the Seventeenth Century," *Journal of the History of Ideas* 3:1 & 2 (1942): pp. 51–73, 190–219, the virtuosi were interested in collecting knowledge not as an avenue to objective truth or law but as a subjective exercise in sensitivity or sensibility. They displayed a curiosity about the marvelous and the incredible which lent itself to traditional interests in religion too (pp. 193–195).

50. Yates, *Giordano Bruno*, p. 233. An English example of both Hermetic influence and the ethic of social utility is Bacon's *New Atlantis*. For fuller discussion, see Chapter 4.

51. Hayden V. White, "The Tasks of Intellectual History," *Monist* 53(1969): p. 606.

52. Ronald Levao, *Renaissance Minds and Their Fictions: Cusanus, Sidney, Shakespeare* (Berkeley: University of California Press, 1985), p. 256. Levao is but one of a number of scholars featuring the critical perspective in relation to ideology evident in English Renaissance literature in general, and in Shakespeare in particular. For example, Jonathan Dollimore in *Radical Tragedy* argues that "Jacobean tragedy discloses ideology as misrepresentation; it interrogates ideology from within . . . offering alternative ways of understanding social and political process" (p. 8). And Stephen Greenblatt suggests in "Invisible Bullets: Renaissance Authority and its Subversion," *Glyph 8: Johns Hopkins Textual Studies* (Baltimore: Johns Hopkins University Press, 1981), that "Shakespeare's plays are centrally and repeatedly concerned with the production and containment of subversion and disorder" (p. 53). These studies, as do others, offer exciting readings of particular literary texts, but they concentrate, to a fault, I believe, in uncovering the self-contestatory quality of each text and fail to examine cultural contestation across various texts—contextually. Perhaps this derives from their emphasis on creative literature per se, but it results in ahistorical interpretations which reduce history and meaning to self-contradiction.

53. Francis Fergusson, *The Idea of a Theater: A Study of Ten Plays, The Art of Drama in Changing Perspective* (Garden City, N.Y.: Doubleday, 1953), p. 153.

54. I offer Hobbes as a focal point (as I do Shakespeare) because I believe that he most particularly crystallizes, in his works, theoretical positions argued by a range of contemporary political thinkers. Rather than anomalous, Hobbes' work is arguably

paradigmatic in its impact. In a similar way, Richard Tuck maintains in *Natural Rights Theories: Their Origin and Development* (Cambridge: Cambridge University Press, 1979), that Hobbes must be viewed within a context "of fundamentally like-minded theorists," particularly the milieu of Selden and his followers (p. 4).

Chapter 1

1. As I have shown already, literary critics especially, but some others too, have begun to correct this interpretation. The goal must be, I believe, to reexamine the character of the Tudor idea of order, not merely to deny its cultural validity. Leaning too far in this latter direction is Moretti who quotes this passage and concludes that "this demonstrates . . . not the force of Elizabethan ideology in Shakespeare, but rather its weakness." See "Tragic Form," p. 36.

2. Remembering our discussion of paradigms in the Introduction, it may be useful to note another of Pocock's arguments. "A paradigm is less a happening than an institution: a reference point within the structure of consciousness, stable or durable enough to be used at more than one moment, and so by more than one actor in more than one way. It is this which makes theory possible, for we have defined theory as the explication of diverse functions and meanings of paradigms." See "On the Non-Revolutionary Character of Paradigms" in *Politics, Language and Time*, p. 280.

3. Thomas Starkey, *England in the Reign of King Henry the Eighth: Dialogue*, ed. J. M. Cowper (London: EETS, 1878), p. 101. Thomas F. Mayer suggests that "the *Dialogue* represents one of the first attempts to blend continental humanism . . . with native English traditions in the creation of a theoretical justification for what Starkey called a 'mixd state,'" in "Faction and Ideology: Thomas Starkey's *Dialogue*, " *Historical Journal* 28:1 (1985): p. 1.

4. Thomas Starkey, *Exhortation to the People Instructing Them to Unity and Obedience* (London, 1536), p. 8.

5. John Aylmer, *An Harborowe for Faithful and Trew Servants* (London, 1559), Sig. H2–3.

6. See F. Le Van Baumer, *The Early Tudor Theory of Kingship* (New Haven: Yale University Press, 1940), p. 90.

7. See "An Exhortation concerning Good Order and Obedience to Rulers and Magistrates," *Homilies* (London, 1817), p. 99, and "An Homily against Disobedience and Willful Rebellion," p. 518.

8. Richard Crompton, *A Short declaration of the end of Traytors, and of false Conspirators against the State, & of the duetie of Subjectes to theyr Sovereigne Governour* . . . (London, 1587), Sig. B2, B4.

9. From the 1550s Puritans and Catholics displayed less respect for this "authority." While Starkey and others had discussed resistance to tyrants in the early years of the century, only by such men as John Knox and John Ponet was this idea argued fully. See Baumer, *Early Tudor Theory*, p. 90.

10. A thorough reading of Arthur Lovejoy's classic *The Great Chain of Being* (New York: Harper & Row, 1960) is essential for an empathetic understanding of medieval and early modern European cosmology.

11. *Homilies*, p. 95.

12. Ibid.

13. Sir Thomas Elyot, *The Book named The Governor*, ed. S. E. Lehmberg (London: J. M. Dent, 1962), p. 2.

14. Ibid., p. 3.

15. Aylmer, *Harborowe*, Sig. C3.
16. Ibid., Sig. M2.
17. Ibid.
18. Ibid., Sig. M3.
19. Starkey, *Exhortation*, p. 2.
20. Ibid.
21. Aylmer, *Harborowe*, Sig. M2.
22. Richard Morison, *A Remedy for Sedition* (London, 1536), p. 3.
23. John Ponet, *A Short Treatise of politike power* . . . (Strasburg, 1556), p. cviii.
24. George Whetstone, *A Mirrour for Magistrates of Citties* (London, 1584), p. 7. See Shakespeare's *Coriolanus* for an interesting and humorous account of the common people as the big toe of the commonwealth.
25. Aylmer, *Harborowe*, Sig. Q4.
26. See this chapter's earlier brief discussion of resistance to tyranny.
27. Aylmer, *Harborowe*, Sig. D4.
28. Ibid., Sig. E3.
29. Ibid.
30. Norton and Sackville, *Gorboduc*, in *Medieval and Tudor Drama*, p. 452.
31. Ibid., p. 419.
32. See E. W. Talbert, *The Problem of Order: Elizabethan Political Commonplaces and an Example of Shakespeare's Art* (Chapel Hill: University of North Carolina Press, 1962).
33. Starkey, *Dialogue*, p. 46.
34. Sir Thomas Smith, *De Republica Anglorum*, ed. L. Alston (Cambridge: Cambridge University Press, 1906), p. 63.
35. Ibid., p. 49.
36. For an introduction to the history of the concept of representation, see Hanna Fenichel Pitkin, *The Concept of Representation* (Berkeley: University of California Press, 1967). In an etymological appendix Pitkin maintains that in England not "until the sixteenth century do we find an example of 'represent' meaning 'to take or fill the place of another (person), substitute for'; and not until 1595 is there an example of representing as 'acting for someone as his authorized agent or deputy'"(p. 243). Pitkin further claims that Smith's *De Republica Anglorum* was "one of the earliest known applications of the English word 'represent' to Parliament." But Smith does not apply the word to members of Parliament (p. 246).
37. Christopher Morris, *Political Thought in England: Tyndale to Hooker* (Oxford University Press, 1953), p. 1.
38. Jones, *Commonwealth*, p. 216.
39. Elyot, *The Book*, p. 1.
40. Aylmer, *Harborowe*, Sig. Cl, D2.
41. J. W. Allen, *A History of Political Thought in the Sixteenth Century* (London: Methuen, 1941), p. 131.
42. *Homilies*, pp. 548, 551.
43. *Politique Discourses, treating of the differences and inequalities of Vocations, as well Publique as Private*, trans. Ægremont Ratcliffe (London, 1578), Sig. Aii.
44. Ibid., Sig. Aiv.
45. *Homilies*, p. 98.
46. Whether Sidney would have continued to write poetry if he had lived to a mature age is, obviously, an unanswerable question. We know that Fulke Greville, his close friend, tried valiantly to balance his life with service to the monarch and literary ambition. Only a few generations later, however, Sackville's "restoration" descendent, Charles, the crony of

Charles Stuart, confronted little social pressure to give up his private desires in favor of a more participatory role in the government. For a fascinating look at this subject see Stephen Greenblatt, *Renaissance Self-Fashioning*, especially the chapters on More and Wyatt. For Sidney, see Ronald Levao, *Renaissance Minds*, for Greville, see Ronald A. Rebholz, *The Life of Fulke Greville, First Lord Brooke* (Oxford: Clarendon Press, 1971), Joan Rees, *Fulke Greville, Lord Brooke, 1554–1628: A Critical Biography* (London: Routledge & Kegan Paul, 1971), and Jonathan Dollimore, *Radical Tragedy*. And for a good study of the political nature of poetry at this time, see David Norbrook, *Poetry and Politics in the English Renaissance* (London: Routledge & Kegan Paul, 1984). Garry Waller, *English Poetry of the Sixteenth Century* (London: Longman, 1986), pays close attention to recent trends in literary criticism and situates poets and their poetry amidst the pressures of courtly life and politics.

47. Starkey, *Dialogue*, p. 10.

48. Morison, *Remedy*, p. 5.

49. Starkey, *Dialogue*, p. 15. Here and throughout civil law should be understood in a general context and not only in its specific meaning which differentiates it from common law and statute law.

50. See Baumer, *Early Tudor Theory*, pp. 6–10.

51. Richard Hooker, *Hooker's Ecclesiastical Polity, Book VIII* (from the Dublin Ms.), introduction by R. A. Houk (New York: Columbia University Press, 1931), p. 178.

52. Baumer, *Early Tudor Theory*, p. 157.

53. Quoted in Baumer, *Early Tudor Theory*, p. 76. Baumer argues from this that Henrician pamphleteers made the king in Parliament an absolute sovereign and established the prerogative of the state for all spiritual concerns. He argues, too, that by 1534 the secular sovereign control of the Church that Hooker, and later Hobbes, advocated was firmly established. I find these arguments, and other arguments that view the 1530s as revolutionary years in English political history, misleading. Until Englishmen accepted the fact that laws were created by man to define a specific, nonexternally determined order, there was no real expression of secular sovereignty. The idea of secular sovereignty depended upon an awareness of man's responsibility for his own social order. See Chapters 4 and 5 for a fuller discussion of these points.

54. Ponet, *Treatise*, Sig. Biii.

55. Crompton, *Short declaration*, Sig. E3.

56. Biographies of Henry VIII and of Elizabeth offer a good look at the use monarchs made of ceremony. Lacey Baldwin Smith's are among the very best: *Henry VIII: The Mask of Royalty* (Boston: Houghton Mifflin, 1973), and *Elizabeth Tudor: Portrait of a Queen* (Boston: Little, Brown, 1975). See also David Moore Bergeron, *English Civic Pageantry, 1558–1642* (Columbia: University of South Carolina Press, 1971), and Roy Strong, *Splendor at Court: Renaissance Spectacle and the Theater of Power* (Boston: Houghton Mifflin, 1973), and *The Cult of Elizabeth: Elizabethan Portraiture and Pageantry* (London: Thames and Hudson, 1977).

57. W. Gordon Zeeveld, *The Temper of Shakespeare's Thought* (New Haven: Yale University Press, 1974), p. 16.

58. Ibid.

59. Elyot, *The Book*, p. 120.

60. Ibid., p. 121.

61. Ratcliffe, *Discourses*, in the dedication, Sig. Aii.

62. Aylmer, *Harborowe*, Sig. B2.

63. Elyot, *The Book*, p. 97.

64. Starkey, *Dialogue*, p. 57.

65. Ibid., p. 66.

66. Ibid., p. 2.

67. Ibid., p. 5.

68. Lacey Baldwin Smith, *Treason in Tudor England*, p. 92. Decorum, "next to piety . . . was the cardinal goal of Tudor education," according to Smith (p. 86), and it contextually institutionalized the role-playing self in Tudor England. Decorum indentified rules associated with roles. "Above all else," according to Smith, "decorum was a style of thinking, a system of cognitive organization which maintained that 'ill doings breed ill thinkings' and from 'corrupted manners spring perverted judgments'" (p. 88). Stephen Greenblatt examines the role-playing feature of "self" and "identity" more fully in *Renaissance Self-Fashioning*. For a good introduction to the concept of decorum, see T. McAlindon, *Shakespeare and Decorum* (New York: Harper & Row, 1973). According to McAlindon, decorum is linked to "opportunity" and thus regulates convenient and proper action—in time and place (p. 10). And for an interesting recent examination of decorum in the context of the self's struggle to solve the problem of representation, see Jean-Christophe Agnew, *Worlds Apart: The Market and the Theater in Anglo-American Thought, 1550–1750* (Cambridge: Cambridge University Press, 1986).

69. See Chapter 2.

70. Ponet, *Treatise*, Sig. Dvii.

71. Ratcliffe, *Discourses*, Sig. Aiii.

72. William Baldwin, "Preface," in *Mirror for Magistrates*, ed. Lily B. Campbell (Cambridge: University Press, 1938), p. 64.

73. Ibid.

74. George Whetstone, *Aurelia. The Paragon of pleasure and Princely delights* (London: 1593), Sig. C1.

75. Ibid.

76. Ibid., Sig. E3. Lawrence Stone's analysis of the relationship between the rise of affective individualism and the changing nature of social institutions in early modern England broadens the perspective here, and can be seen to corroborate, in many ways, my general thesis about historical psychology and social meaning. See *The Family, Sex and Marriage in England 1500–1800.*

77. Elyot, *The Book*, p. 23.

78. Indeed, Stephen Greenblatt claims that as we move "from Spencer to Marlowe to Shakespeare, we move toward a heightened investment of professional identity in artistic creation." *Renaissance Self-Fashioning*, p. 161.

79. Ibid. In his discussion of Wyatt, in *Renaissance Self-Fashioning*, Greenblatt admits that Wyatt was an imaginative and innovative poet, interested in form, but he warns that it is anachronistic to suggest that Renaissance poetic technique developed "out of a disinterested aesthetic concern for form and apart from both personal interests and the general interests of the culture as a whole." He concludes that Wyatt's "poetry is, in effect, a species of conduct" (p. 136).

80. Commerce and trade, for example, were unacceptable activities by the standards of Tudor order theory. In Chapter 4 I discuss the history of merchant propaganda which tried to overcome social ostracism. The merchants succeeded when they were able to picture trade as socially purposive. Everything was educative, in Tudor England, and according to Smith, education "was a holistic enterprise . . . and no subject, not even arithmetic, was regarded as being devoid of moral meaning." *Treason in Tudor England*, p. 79.

81. Elyot, *The Book*, p. 78.

82. Ibid., p. 79.

83. Ibid., pp. 80–81.

84. Starkey, *Dialogue*, p. 32.
85. Ibid., p. 38.
86. Ibid., p. 4. Man, too, was a purposive agent, not an end in himself.
87. Elyot, *The Book*, p. 163.
88. Starkey, *Dialogue*, p. 12.
89. Starkey, *Exhortation*, p. 6.
90. Ibid., p. 7.
91. Robert Hoopes, *Right Reason in the English Renaissance* (Cambridge: Harvard University Press, 1962), p. 11.
92. Robert Mason, *Reasons Monarchie* (London, 1602), pp. 79–80.
93. Robert Mason, *A Mirrour For Merchants* (London, 1609), pp. 79–80.
94. Hoopes, *Right Reason*, p. 14.
95. Smith, *Treason in Tudor England*, p. 87.
96. Ibid., p. 86.
97. Greenblatt, *Renaissance Self-Fashioning*, p. 162.
98. Sir Thomas Wilson, *The Rule of Reason conteinyng the arte of Logique* (London, 1551), Sig. Eiiii.
99. Cf. Agnew, *Worlds Apart*, p. 93.
100. Mason, *Reasons*, pp. 3–4.
101. Ibid., p. 5.
102. Ibid., introduction.
103. Starkey, *Dialogue*, p. 52.
104. Ibid., p. 165.
105. Elyot, *The Book*, p. 51.
106. Tillyard, *Elizabethan World Picture*, p. 73.
107. Talbert, *Problem of Order*, p. 121.
108. Elyot, *The Book*, p. 5.
109. Ibid., p. 237.
110. Ibid., p. 209.
111. Ibid., p. 200.
112. Ibid., p. 209.
113. Tillyard, *Elizabethan World Picture*, p. 75.
114. Elyot, *The Book*, p. 209.
115. Starkey, *Exhortation*, p. 5.
116. Ratcliffe, *Discourses*, p. 64.
117. Aylmer, *Harborowe*, Sig. A3.
118. Elyot, *The Book*, p. 208.
119. Ibid., p. 207.
120. George Cavendish, *Metrical Visions* (London, 1825), p. 6.
121. Ibid., p. 16.
122. Thomas Sackville, "Induction," in *Mirror for Magistrates*, p. 300.
123. Baumer, *Early Tudor Theory*, p. 125.
124. Elyot, *The Book*, p. 165.
125. Ratcliffe, *Discourses*, Sig. Aiii.
126. Ibid., Sig. Aii–iii.
127. Jones, *Commonwealth*, p. 226.
128. Walzer, *Revolution*, p. 8.
129. Ibid., p. 6.
130. Ibid., p. 9. See also in this context, Arthur B. Ferguson, *Clio Unbound: Perception of the Social and Cultural Past in Renaissance England* (Durham, N.C.: Duke University Press, 1979). Ferguson documents the growing awareness of social change in

Tudor England and identifies its importance for a developing historical and political consciousness; see especially pp. xiii, 430–431.

131. Thomas Hobbes, *Leviathan*, ed. M. Oakeshott (Oxford: Basil Blackwell, 1946), p. 5. (This chapter introduces Hobbes' political thought as expressive of the new idea of order. For a fuller analysis of Hobbes' work, see Chapter 5.)

132. Walzer, *Revolution*, p. 160. Walzer claims: "The changing nature of the political world was, however, paralleled by changes in the conception of the cosmos. . . . These changes in the view of God and his universe had many sources. Calvinism was among the most important." But again I find his emphasis on Calvinism misleading if we are to understand what basic values about man, society, God, and the cosmos characterized the overall change in consciousness that marked meaning redefinition and an acceptance of order based upon the political, secular state after the mid-seventeenth century. Walzer tries to find a reason for the dissolution of the traditional hierarchical view of things in the Puritans' ubiquitous involvement in Satan, sin, and the Fall; Puritans "produced descriptions of chaos which sounded very much like Hobbes' view of nature. And if chaos were natural, there was no great chain," he argues (p. 159). But here again he misses the point. Puritan politics reacted against this chaos; Hobbes and others who articulated secular sovereignty defined a new order and new existential meaning that was based upon it.

133. I am not suggesting that the history of political consciousness, or even of developing self-identity in England, needs to be separated from an understanding of the ramifications of the political and social crises of seventeenth-century England. Indeed, in the chapters that follow I hope to suggest that early seventeenth-century England demonstrated a manifestly dynamic relationship between consciousness and historical praxis. And it is just such a close relationship that enabled the English to define an ordered, secular, and modern society in the years after 1660. What I am stressing here, however, is that we cannot simply ascribe these conceptual changes to social causes. I eschew both the materialist and the idealist explanations of history in the hope of describing the change in historical and social meaning that affected both mind and society so categorically during these years.

134. Thomas Hobbes, *Elements of Law Natural and Politic*, 2nd ed., ed. Ferdinand Tonnies, with new introduction by M. M. Goldsmith (London: Frank Cass, 1969), 14:4. Part 1.

135. Ibid., 14:10–14; 12. Part 1. Chapter 5 deals more fully with Hobbes' struggle with self-negation. Chapter 2 argues that Shakespeare struggles to transcend the necessity of self-negation in his major tragedies and attempts to posit a positive order in the late plays.

136. In "Conquest and Consent: Thomas Hobbes and the Engagement Controversy," in *The Interregnum: The Quest for Settlement 1646–1660*, ed. G. E. Aylmer, (Hamden, Conn.: Archon Books, 1972), p. 91, Quentin Skinner deals clearly with a specific manifestation of these political issues in the last years of the 1640s and identifies various minor theorists who articulated similar positions.Compare Glenn Burgess, "Usurpation, Obligation and Obedience in the Thought of the Engagement Controversy," *Historical Journal* 29:3 (1986): pp. 515–536, with Skinner's essay. Burgess maintains that Skinner overemphasizes the role of independent secular arguments in this controversy and minimizes the role of providentialism.

137. Compare Richard Tuck's discussion of Ascham in *Natural Rights Theories*, pp. 152–154. Tuck emphasizes Ascham's radical rights theories which "could have been endorsed by any of the levellers" (p. 154).

138. From "A Combat Between Two Seconds," quoted by Skinner, *Interregnum*, p. 88.

139. Quoted in ibid.

140. Quoted in ibid., p. 90.

141. Quoted in ibid., p. 89.

142. Quoted in ibid., p. 83.

143. Ibid.

144. Henry Parker, *The True Grounds of Ecclesiastical Regiment* (London, 1641), p. 18.

145. Henry Parker, *Observations upon some of his Majesties late Answers & Expresses*, 2nd ed. (London, 1642), p. 1. See Chapter 5 for a fuller analysis of Parker, and other writers who openly debated with him, such as Hunton, Filmer, and Digges.

146. Ibid. The projection of this inherent power by man is a creative act which contracts political society. Parker here resembles Hobbes and often seems close to articulating a theory of full secular sovereignty. But the maintaining of "inherent power" is confusing and ambiguous. For Parker's idea of secular sovereignty and his likeness to Hobbes, see Chapter 5.

147. Parker, *True Grounds*, p. 25.

148. Ibid., p. 19.

149. Ibid., p. 24.

150. M. A. Judson and W. K. Jordan argue that Parker was, indeed an exponent of a modern type of secular sovereignty. I counter their arguments in Chapter 5. It appears, however, in the few quotations already cited, that Parker makes use of God's "goodness," the idea of the infinite in the one. See Judson's "Henry Parker and the Theory of Parliamentary Sovereignty," in *Essays in History and Political Theory* (Cambridge: Harvard University Press, 1936), and *The Crisis of the Constitution* (New Brunswick, N.J.: Rutgers University Press, 1949), and Jordan's *Men of Substance* (Chicago: University of Chicago Press, 1942).

151. Henry Parker, *Jus Populi* (London, 1644), p. 3.

152. Quentin Skinner concludes his essay on the engagement controversy by stating that Hobbes alone contributed to the discussion "about the rights of *de facto* powers" without recourse to "invocations of God's providence." "It was in his predication of his political system upon a comprehensive account of man's political nature, and in his unique emancipation from the confines of the providentialist vocabulary, that Hobbes made his most original contributions to political theory." Skinner, *Interregnum*, p. 98. I am not willing to accept that herein lies Hobbes "most original contributions to political theory," but I do agree that Hobbes' originality was epistemological more than it was political. Chapters 4 and 5 show that the changes in political philosophy that surfaced in the 1640s were conditioned by changes in the manner of knowing and perceiving the world as much as by reactions to social crises.

153. Parker, *Jus Populi*, p. 7.

154. Ibid., p. 8.

155. Parker, *Observations*, p. 15. Parker is concerned here specifically with the issue of liberty versus prerogative: for the relationship between this political issue and political theory, see Chapters 4 and 5.

156. Thomas Hobbes, *De Cive* in *The English Works of Thomas Hobbes*, ed. Sir William Molesworth (London, John Bohn, 1839–1845), vol. 2.

157. William R. Lund, "Tragedy and Education in the State of Nature: Hobbes on Time and the Will," *Journal of the History of Ideas* 48:3 (1987), p. 394.

158. For a fuller discussion of this epistemological point see J. W. N. Watkins, *Hobbes's System of Ideas* (London: Hutchinson, 1965), especially Ch. 4. Cf. Tuck's discussion of Hobbes in *Natural Rights Theories*.

159. Hobbes, *Leviathan*, p. 141.

160. Sir Walter Raleigh, *The Works of Sir Walter Ralegh, Political, Commercial, and Philosophical* (London, 1751), p. 2. See also Ferguson, *Clio*, for an account of the growing sense in Renaissance England of the relativity of customs, institutions, and values.

161. In William Blake's poem *Tyger Tyger*, this dual sense of "Frame" is magestically expressed. In its dialectical character "framing" suggests a political counterpart to Greenblatt's "fashioning."

162. Raleigh, *Works*, p. 1.

163. Parker, *Jus Populi*, p. 29.

164. Ibid., p. 35.

165. Henry Parker, *The Case of Shipmony* (London, 1640), pp. 3–4.

166. Ibid.

167. Parker, *Jus Populi*, p. 17.

168. Ibid., pp. 17–19.

169. Ibid., p. 18, emphasis mine. Pitkin tells us that "both the formalistic sense of representation and its substantive correlate 'acting for' emerged during this period [middle decades of the seventeenth century], apparently by way of the idea that Parliament represented the whole realm, which in turn began in the notion of a mystic or symbolic 'standing for.'" *The Concept of Representation*, p. 250.

170. Psychoanalysis teaches about the analogous organization of the sexual body. For some considerations gleaned from N. O. Brown and from Eric Voegelin's discussion of Fortescue, see the Conclusion of this book.

171. Hobbes, *Elements of Law*, 19:5. Part 1.

172. Ibid., 19:10. In *De Cive* and in *Leviathan*, Hobbes' arguments tend to support the view that subjects contract among themselves only, thus denying the sovereign any original authority.

173. Hobbes, *Leviathan*, pp. 112–113. Pitkin generalizes about Hobbes' place in the etymological development of "represent" in *Concept of Representation*, p. 250, and Agnew relates Hobbes' "dynamic" usage of "representation," his sense of "surrogate agency," with the theatrical metaphor so common in *Leviathan*, in *Worlds Apart*, p. 102.

174. Hobbes, *De Cive*, 6:19.

175. Raleigh, *Works*, p. 140.

176. Hobbes, *Leviathan*, p. 83.

177. Hobbes, *Elements of Law*, 10:1. Part 2.

178. Hobbes, *De Cive*, 3:33. In *Natural Rights Theories* Tuck discusses the relationship between obligation and rational certainty. Laws of nature could be propounded as deductive conclusions. While Hobbes had a "modern" theory of probability, claims Tuck, "he refused to apply it to his political theory." Men contracted and obligated themselves based not upon calculated probable benefit, but upon certainty. Only certainty rationally rules out renouncing rights of resistance (p. 128).

179. Hobbes, *Leviathan*, pp. 31–32.

180. Hobbes, *De Cive*, 3:29.

181. Hobbes, *Leviathan*, p. 174.

182. Ibid., p. 176.

183. Michael Oakeshott, *Rationalism in Politics* (London: Methuen, 1962), p. 280.

184. Watkins, *Hobbes's System*, pp. 103–110. Note that "motive" should be understood here in its full significance as motion as well as in the specific psychological sense.

185. Oakeshott, *Rationalism*, p. 258.

186. Hobbes, *De Cive*, 1:15, 2:1n.

187. Ibid.

188. Ibid., introduction p. xvi. The analogy to the psychosexual development of children as they "mature" is enlightening here. Hobbes, along with Montaigne, was truly a pioneer in modern psychology.

189. Hobbes, *Leviathan*, pp. 25–26.

190. Ibid., p. 19.

191. Hobbes, *Elements of Law*, 5:6. Part 1.
192. Hobbes, *Leviathan*, p. 29.
193. Ibid., p. 17.
194. Ibid., p. 32.
195. Ibid., p. 30.
196. Ibid., p. 57.
197. For an interesting interpretation of Hobbes in a similar but more blatantly Marxist vein, see C. B. Macpherson, *The Political Theory of Possessive Individualism: Hobbes to Locke* (London: Oxford University Press, 1964). Cf. Tuck, *Natural Rights Theories*. And for a discussion of Shakespeare along these lines, see Chapter 2.
198. Hobbes, *Leviathan*, p. 60.
199. Ibid., p. 62.
200. Hobbes, *De Cive*, 3:11.
201. Hobbes, *Elements of Law*, 17:1. Part 1.
202. Hobbes, *Leviathan*, p. 6.
203. Jones, *Commonwealth*, p. 227.
204. Hobbes, *De Cive*, 15:14.
205. Parker, *Jus Populi*, pp. 36, 43.
206. Ibid., p. 57.
207. Hobbes, *De Cive*, 6:11.
208. Hobbes, *Leviathan*, p. 64.
209. Ibid., p. 63.

Chapter 2

1. Quoted in Lovejoy, *The Great Chain of Being*, p. 19.
2. All references to Shakespeare's plays in this chapter are from *The New Shakespeare*, ed. John Dover Wilson, Sir Arthur Quiller-Couch, et al. (Cambridge: Cambridge University Press, 1931–1971) except for references to *A Midsummer Night's Dream* and *As You Like It*, which come from the Signet Classic Shakespeare, gen. ed. Sylvan Barnet; respectively edited by Wolfgang Clemen and Albert Gilman (New York: New American Library, 1963). All references to Marlowe's plays in this chapter are from the Everyman's Library editions of *Plays and Poems*, ed. M. R. Ridley (London: J. M. Dent, 1973). The exceptions are the references to *Doctor Faustus*, which come from *The Complete Plays*, ed. Irving Ribner (New York: Odyssey Press, 1963).
3. Tillyard, *Elizabethan World Picture*, p. 9.
4. Ibid., p. 9, I am singling Tillyard out only because he so cogently expressed these views. An immense variety of corroborating references are available.
5. Ibid., p. 4.
6. C. L. Barber, *Shakespeare's Festive Comedy: A Study of Dramatic Form and Its Relation to Social Custom* (Cleveland: World, 1968), pp. 221, 15. Some other interesting studies that focus upon changes in relation to the "Elizabethan World Picture" are Richard Dollimore, *Radical Tragedy*; J. W. Lever, *The Tragedy of State* (London: Methuen, 1971); Norman Rabkin, *Shakespeare and the Common Understanding* (New York: Free Press, 1967); Terry Eagleton, *William Shakespeare* (Oxford: Basil Blackwell, 1986); the various works of Stephen Greenblatt, and those of Louis Adrian Montrose, especially "The Purpose of Playing: Reflections on a Shakespearean Anthropology," *Helios* n.s. 7:2 (1980): pp. 51–74. In this essay Montrose treats the "functional" character of Shakespeare's theater. "In a society in which the dominant social institutions and cultural forms are predicated upon an ideology of unchanging order and absolute obedience, a new kind of

commercial entertainment that is still imbued with the heritage of suppressed popular and religious traditions can come to serve a vital social function . . . Shakespeare's theatre articulates—and thereby helps its heterogeneous audience of social actors to adjust to and to control—the ambiguities and conflicts, the hardships and opportunities, arising from the ideologically anomalous realities of change" (p. 64). The general tenor of this interpretation is useful, but the overtly "functionalist" reduction, and the reductive view of Tudor ideology, the overemphasis on its practical coerciveness, should be questioned. They lead to overdetermining antistructural, antiauthoritarian theatrical characteristics. In every walk of life, and in all discursive forms, Elizabethans were behaving "marginally" and were producing marginal experiences while *not* turning their "world upside down." The key here, I believe, is that Montrose (and a number of other self-identified new literary historicists) sees order and change as inhabiting and deriving from separable worlds. Change is not necessarily social discontinuity and both change and order are the products of continuous social action. Montrose's claim that theater functioned as an undermining (not merely an interrogating) agent rests on the point that it refuted "not everyday life at all, but rather an idealized or coercive hierarchical model" (p. 65). But does this not (using Montrose's "functionalist" perspective) suggest the theater's social redundancy, for surely everyday life is always already reconstituting social reality by undermining the model of order and reconceptualizing this process. The idea of order does not coercively limit everyday life if the latter does not re-produce it socially. From this perspective, theater participates in articulating the problem of order and helps produce and "naturalize" a new concept of everyday life rather than merely undermining coercive ideology.

7. Levao, *Renaissance Minds*, p. 293. Three other recent works that offer alternative yet related views are Jane Donawerth, *Shakespeare and the Sixteenth-Century Study of Language* (Chicago: University of Illinois Press, 1984); James P. Driscoll, *Identity in Shakespearean Drama* (Lewisburg, Penn.: Bucknell University Press, 1983); and Anne Ferry, *The "Inward" Language: Sonnets of Wyatt, Sidney, Shakespeare, Donne* (Chicago: University of Chicago Press, 1983). Of special note is Driscoll's *Identity in Shakespearean Drama*, which treats "identity" more specifically and reductively than I do—emphasizing a Jungian reading. Nevertheless, in many places Driscoll's interpretations can be read as complementary to mine, even as they suggest opposite conclusions. See especially the fine discussions of Falstaff and Prospero in his chapters 1 and 7.

8. Irving Ribner, "Marlowe and Shakespeare," *Shakespeare Quarterly* 15:2 (1964): p. 44.

9. Charles S. Masinton, *Christopher Marlowe's Tragic Vision* (Athens: University of Ohio Press, 1972), p. 12.

10. Norton and Sackville, *Gorboduc*, in *Medieval Tudor Drama*, p. 414. All citations are from this edition.

11. For a discussion of *Gorboduc* as more paradoxical and self-contradictory than I have suggested it is, see Norman Rabkin "Stumbling Toward Tragedy," in *Shakespeare's "Rough Magic": Renaissance Essays in Honor of C. L. Barber*, eds. Peter Erickson and Coppélia Kahn (Newark, N.J.: University of Delaware Press, 1985), pp. 28–49).

12. Masinton, *Christopher Marlowe's Tragic Vision*, p. 8.

13. For an interesting psychological discussion of this point, and for a fine interpretation of the entire play, see C. L. Barber, "The Form of Faustus' Fortunes Good or Bad," in *Shakespeare's Contemporaries: Modern Studies in English Renaissance Drama*, eds. Max Bluestone and Norman Rabkin (Englewood Cliffs, N.J.: Prentice-Hall, 1970).

14. The motif of sacrifice-salvation runs through Shakespeare's tragedies. As discussed later, Caesar, Hamlet, Cordelia, and Coriolanus are sacrificed in magical attempts to exorcise the seeds of social disorder.

15. Barber, "Faustus' Fortune," p. 110.

16. Lovejoy, *The Great Chain of Being*, p. 320.

17. Ibid., p. 321.

18. Ibid.

19. D. J. Palmer, "Magic and Poetry in 'Doctor Faustus,'" *Critical Quarterly* 6:1 (1964): p. 62.

20. Barber, "Faustus' Fortune," p. 97.

21. Ibid., p. 98.

22. Ibid., p. 107.

23. Erich Heller, "Faust's Damnation: The Morality of Knowledge," in *The Artist's Journey into the Interior and Other Essays* (New York: Random House, 1965), pp. 17–18.

24. From a different perspective, Jonathan Dollimore notes that the question of whether or not Faustus repents "is less important than the fact that he is located on the axes of contradictions which cripple and finally destroy him." See *Radical Tragedy*, p. 112.

25. Masinton, *Christopher Marlowe's Tragic Vision*, p. 77.

26. Harry Levin, *Christopher Marlowe: The Overreacher* (London: Faber & Faber, 1952), p. 188.

27. In his *Renaissance Self-Fashioning*, Greenblatt depicts Marlovian "self" as differently constituted than others he discusses, "achieved through a subversive identification with the alien." In opposition is the "fashioned" self, "achieved through an attack upon something perceived as alien and threatening" (p. 203). In conceptualizing this way, Greenblatt focuses the difference between what I suggest is self-definition and self-fashioning; between the theme of my book and his. My aim is to elaborate con-textually this Marlovian identity; Greenblatt, on the other hand, notes its textually anomalous presence.

28. Karl Jaspers, *Tragedy Is Not Enough*, trans. Harold A.T. Reiche, Harry T. Moore, and Karl W. Deutsch (Boston: Beacon Press, 1952), p. 49.

29. For much that informs my specific understanding of Shakespeare's early comedies and the *Richard II–Henry IV* cycle, I am indebted to C. L. Barber's *Shakespeare's Festive Comedy*.

30. Ibid., pp. 159–160.

31. Ibid., p. 148.

32. In *William Shakespeare*, Eagleton expands on the possibilities of such an interpretation. "If the hallucinated mismatchings of the forest are framed by the sober contract of marriage, that in turn is framed by self-conscious theatrical illusion. In foregrounding its own fictional character, the play wards off the disorderly desire it has itself unleashed" (p. 26). And in "The Purpose of Playing," Montrose elaborates the textual self-contestation implicated in the festive comedies. Their happy endings, he says, "can be viewed as symbolic assimilations of potential disorder by a normative system." But these plays "also embody serious and partially successful challenges to the patriarchal order of the family and the polity" (p. 67).

33. I appreciate that disguises were an accepted convention in the theater of the time but I have often wondered why they were accepted. I believe that the answer may lie in the yet undeveloped perceptive psychology of the time which Lucien Febvre among others has noted. See *The Problem of Unbelief in the Sixteenth Century: The Religion of Rabelais*, trans. Beatrice Gottlieb (Cambridge, Mass.: Harvard University Press, 1982). The development of a fuller perceptive psychology must accompany the development of the consciousness of self and other that I am concerned with here. More can be done with this in relation to this popular theatrical convention, I believe. From a different and interesting interpretive vantage, Kent T. van den Berg, *Playhouse and Cosmos: Shakespearean Theater as Metaphor* (Newark, N.J.: University of Delaware Press, 1985), relates disguise to "theatrical artifice" which for audience as for characters "is a means to truth." In *As You Like It*, he argues, this is particularly clear. "While Rosalind's disguise gives self-conscious

emphasis to theatrical artifice, it also functions as a metaphor of her response to love, disclosing in the outward form of theatrical counterfeiting the inner reality of her character" (p. 11).

34. McAlindon discusses this point from the perspective of "decorum," while criticizing Tillyard and others for identifying Richard particularly with "ceremony and propriety" and Bolingbroke with "pragmatism," in *Shakespeare and Decorum* (pp. 34–35).

35. Again McAlindon's discussion of *Richard II* in *Shakespeare and Decorum* broadens this perspective.

36. Barber, *Festive Comedy*, p. 195.

37. Eagleton traces "linguistic inflation" and free-floating symbols, signs as fetishes, in Shakespeare, with particular attention to *Richard II* and *Henry IV* in *William Shakespeare* (pp. 8–17).

38. For an interesting discussion of this final scene along these lines, see Barber, *Festive Comedy*, pp. 217–221.

39. Northrup Frye, *Fools of Time: Studies in Shakespearean Tragedy* (Toronto: University of Toronto Press, 1967), p. 5.

40. Ibid., p. 6. By such an understanding of the nature of the tragic process one can agree with Frye that while tragedy is existential, existentialism is post-tragic, p. 4. One can see Hobbes' relationship to Shakespeare in this light.

41. Ibid., p. 35.

42. Ibid., p. 29.

43. Norman Rabkin elucidates this theme, which borrows from Niels Bohr's theory of "complementarity," in *Shakespeare and the Common Understanding*. Rabkin suggests that Shakespeare's vision views experience as possessing radically inimical yet equally viable meanings; see his ch. 1, pp. 10, 24.

44. R. J. Kaufmann and Clifford J. Ronan, "Shakespeare's *Julius Caesar:* An Apollonian and Comparative Reading," *Comparative Drama* 4 (1970): p. 24.

45. With an ironic twist to Brutus' use of names and public reputation, Antony conjures a spell against Brutus when he addresses the people with his repetitive use of Brutus' name and "honourable man."

46. Remember, *Julius Caesar* begins with the celebration of the Lupercalia. For a fascinating account of the relationship between politics and wolf-imagery see Norman O. Brown, "XV. Kal. Mart. (February 15) LUPERCALIA" *New Literary History* 4:3 (Spring 1973): pp. 541–556.

47. Peter S. Anderson, "Shakespeare's *Caesar:* The Language of Sacrifice," *Comparative Drama* 3 (1969): p. 7.

48. For two good accounts of the affinities between Caesar and Brutus, sacrificed and sacrificer, see Anderson, "Shakespeare's *Caesar*" and Rabkin, *Shakespeare and the Common Understanding*, pp. 105–119.

49. R. J. Kaufmann, "Ceremonies for Chaos: The Status of 'Troilus and Cressida,'" *English Literary History* 32:2 (1965): p. 142.

50. Terence Eagleton, *Shakespeare and Society: Critical Studies in Shakespearean Drama* (London: Chatto & Windus, 1970), p. 146. In *William Shakespeare*, Eagleton specifies the implications of this view. In *Troilus and Cressida*, he argues, social relations "are not simply the medium within which an individual may choose to express his already well-formed identity, but the very discourse which constitutes the self. . . . It would appear that the self, like signs and values, has no choice between being fully, autonomously itself, which is a kind of nothing, and being a shifting cypher wholly dependent on context, which is another kind of nothing" (p. 61). But there is a choice; the one Shakespeare's tragic vision dramatically works through and Hobbes' political philosophy incorporates. Because if social intercourse *is* the very discourse through which identity is constituted, it also

simultaneously produces social meaning and order. And more important, it is a negotiable process, continually confirming and transforming identity and ideology. And it is during specifically transformative moments that discourse—this negotiable process—becomes apparently self-conscious, thus allowing for a historically "meaningful" rejection or negation of both "nothings."

51. Kaufmann, "Ceremonies for Chaos," p. 146.

52. Rabkin, *Shakespeare and the Common Understanding*, p. 53.

53. Erich Heller, "The Tempestuous Retreat," *The Listener* (January 21, 1965): p. 103.

54. Eagleton, *Shakespeare and Society*, p. 56.

55. Ibid., pp. 43–48.

56. Ibid., p. 62. Much of the preceding discussion is indebted to Eagleton's perceptive analysis.

57. Pocock's comments about the linguistic nature of politics in "Languages and Their Implications," can be applied to *Hamlet* particularly and Shakespeare's tragic vision from *Hamlet* to *Coriolanus* generally. "To speak at all is to give some other power over us, and some assert their own power by refusing to speak at all, to speak intelligibly or (so far as this is possible) within any frame of reference they cannot unilaterally prescribe" (p. 24). Hamlet, Lear, and Coriolanus all know, or sense, this. Social discourse is especially precarious when fixed relations are tenuous. When language is inflated, when signs roam free of signifieds, when ceremony and ritual are empty, yet coercive forces, communication is dangerous because everything is open to interpretation. Without a "meaningfully" negotiated social context within which the individual can act, all interaction threatens the self which experiences both its own power and its vulnerability. This experience is what the "Shakespearean Moment" dramatizes.

58. Eagleton, *Shakespeare and Society*, p. 132.

59. Ibid.

60. Ibid., p. 105.

61. Ibid., p. 151.

62. Frye, *Fools of Time*, p. 121.

63. Eagleton, *Shakespeare and Society*, p. 140.

64. Ibid.

65. Ibid., p. 141.

66. I have been presumptuous enough to use the term "imagication" to refer to Prospero's special creative imagination which magically and artfully taps nature's processes. From a Marxist perspective, Eagleton, in *William Shakespeare*, labels the last comedies as "ideological" in that their discursive task "is to naturalize a particular social order, imbuing it with the immutability and inevitability of Nature itself" (p. 93). But such an interpretation denies the "historical" in Shakespeare and ignores the culturally redefined meaning of "nature" that these plays privilege. The social order particularized in *The Tempest*, for instance, results from a negotiated cultural process and is significantly different from ones projected in earlier plays.

67. Frank Kermode, "Introduction" to *The Tempest*, p. xlviii.

68. Eagleton, *Shakespeare and Society*, p. 164.

Chapter 3

1. For an important account of the relationship between a developing historical consciousness and the awareness of social change and variety in Tudor England, see Ferguson's *Clio*.

2. This "identity" must be understood in a commonsense, psychological manner. I am not using identity as a metapsychological structure. For a discussion of "identity" from critical, philosophical, and psychological perspectives see Driscoll, *Identity in Shakespearean Drama*, introduction. For discussions of Tudor identity see Lacey Baldwin Smith, *Treason in Tudor England*, ch. 3, and my analysis in Chapter 1.

3. See George Mosse, *The Struggle for Sovereignty in England from the Reign of Queen Elizabeth to the Petition of Right* (New York: Octagon Books, 1968), p. 27.

4. The conceptual structure for this chapter, as well as for the remaining chapters, is supported by Thomas Kuhn's theories about change as detailed in his *The Structure of Scientific Revolutions*. I have spoken to this in the Introduction. Without belaboring the content of the work with heavy-sounding terminologies, I am trying to show that meaning redefinition may be understood in terms of paradigms and anomalies.

5. For a discussion of these elements which became part of the traditional Renaissance epistemology, see Lovejoy's *Great Chain of Being*. See also Ferguson's *Clio*, for a fresh interpretation of this subject.

6. Starkey, *Exhortation*, p. 1.

7. For a good introduction to the general ideas of the English humanists see Fritz Caspari, *Humanism and the Social Order in Tudor England* (Chicago: University of Chicago Press, 1954).

8. Shakespeare's *The Tempest* and Bacon's *New Atlantis* are two pieces that develop these themes most intensely.

9. Starkey, *Exhortation*, p. 1, and Starkey, *Dialogue*, p. 6.

10. See a discussion of this in L. Alston's introduction to Sir Thomas Smith's *De Republica Anglorum*, p. XXXV.

11. Smith, *De Republica Anglorum*, p. 13.

12. Ibid., p. 33.

13. Ibid., p. 39.

14. Ibid., p. 28.

15. Ibid., p. 28.

16. William Fulbeck, *The Pandectes of the Law of Nations* (London, 1602), p. 19.

17. Ibid., p. 60.

18. Ibid., p. 90.

19. Ibid., p. 85.

20. Francis Bacon, "Certain Observations Made Upon a Libel Published This Present Year, 1592," *The Letters and the Life of Francis Bacon Including all his Occasional Works*, ed. James Spedding (London: Longmans, Green & Co., 1861–1874), 1:154 (hereafter cited as Bacon, *L & L*).

21. Smith, *De Republica Anglorum*, pp. 14–15.

22. Ibid., p. 21.

23. Charles Merbury, *A Briefe Discourse of Royall Monarchies, As of The Best Common Weale* (London, 1581), p. 6.

24. Ibid., p. 7.

25. Ibid., p. 41.

26. Ibid., p. 44.

27. Wilson, *Rule of Reason*, Sig. Eii.

28. Ibid., Sig. Eiii.

29. Ibid.

30. Ponet, *Treatise*, Sig. cvi.

31. Crompton, *Short declaration*, Sig. E3.

32. Ibid., Sig. E4.

33. William Lambard of Lincolnes Inne, Gent., *Eirenarche: or of The office of the Iustices of Peace in Foure Bookes*, 2nd ed. (London, Ralph Newbery, 1592), p. 7.

34. Ibid., p. 9.

35. William Fulbeck, *A Direction or Preparative to the Study of the Lawe* (London, 1600), pp. 2, 8.

36. Ibid., p. 8.

37. Thomas Floyd, *The Picture of a perfit Common wealth* (London, 1600), p. 61.

38. Ibid., pp. 4–5.

39. Ibid., p. 6

40. Ibid., p. 65.

41. George Whetstone, *The English Myrror* (London, 1586), p. 199.

42. Ibid.

43. Floyd, *Picture*, pp. 78, 81.

44. Ratcliffe, *Discourses*, pp. 54–55.

45. Fulbeck, *Direction*, p. 31.

46. Sir Thomas Smith, *A Discourse of the Commonweal of This Realm of England* (London, 1549), ed. Mary Dewar, Folger Shakespeare Library (Charlottesville: University Press of Virginia, 1969), introduction, p. 10.

47. William Cecil, Lord Burghley, "Certain Precepts for the Well Ordering of A Man's Life" (c. 1584), in *Advice to a Son*, ed. L. B. Wright (Ithaca, N.Y.: Cornell University Press, 1962), p. 10.

48. Ibid., pp. 12–13.

49. I will deal with Bacon's political ideas and with his intellectual life in some detail in Chapter 4.

50. Francis Bacon, "On The Controversies of the Church" (1589), in Bacon, *L & L*, 1:92.

51. Ibid., p. 123.

52. From Sir Simonds *D'Ewes Journals*, also quoted in J. E. Neale, *Elizabeth I and Her Parliaments 1559–1581* (London: Jonathan Cape, 1953), p. 49.

53. Quoted in ibid.

54. Floyd, *Picture*, pp. 149–150.

55. Richard Crompton, *The Mansion of Magnanimitie* (London, 1599), Sig. A4. For a clear, historical discussion of the relationship between fame and redemption, see Huizinga's *The Waning of the Middle Ages*. Eric Fromm addresses the relationship between the meaninglessness of life and the aspiration for fame in *Escape from Freedom* (New York: Avon Books, 1941).

56. Crompton, *Mansion*, Sig. A4.

57. Ibid., Sig. E2.

58. Ibid., Sig. F3.

59. Sir Richard Barckley, *A Discourse of the Felicitie of Man* (London, 1598), preface, Sig. A3.

60. Ibid.

61. Ibid.

62. Such a retreat into the confines of comfortable traditions is described by Huizinga in *The Waning of the Middle Ages*. The attempt to deal with anomalous experience, whether internal or external, often leads initially to a rejection of the new and an attempted revival of the traditional. Thomas Kuhn's remarks in *The Structure of Scientific Revolutions* have already been noted. Finally, more subtle attempts at conservative confrontations are often defined during transitional moments, but are in reality unconsciously couched innovations. (see also Chapter 4's discussion of English Hermeticism.) I am suggesting here that the social and internal dynamic that marked this historical epoch can

be understood in these general terms and I am offering certain discussions and themes as microcosmic analogy.

63. Barckley, *Felicitie of Man*, Sig. A7 (preface).

64. Ibid., pp. 131–132.

65. Ibid., p. 455.

66. Ibid., p. 465.

67. Ibid., p. 483.

68. Ibid., p. 504.

69. Ibid.

70. Ibid., p. 569.

71. Ibid., p. 545.

72. Ibid., pp. 545–546.

73. Ibid., p. 525.

74. Without arguing about the connotations of the term *destined*, or about the psychological possibilities of self-fulfillment that disregards "other," I offer here the vision dramatized by Shakespeare from *Troilus and Cressida* to *Coriolanus*. My intention is merely to translate this vision into probable social and political consciousness as we can glean it from various written expressions.

75. All quotations are taken from James Oliphant's abridged edition of *The Educational Writings of Richard Mulcaster* (Glasgow, J. Maclehose and sons, 1903), specifically the first part of Mulcaster's *Elementarie* written about 1582. Oliphant paraphrased the work in "modern English," p. 1.

76. Ibid., p. 7.

77. Ibid., p. 9.

78. Ibid., p. 11. Although Mulcaster emphasized circumstantial evaluation of social and historical conditions, he still maintained a traditional moral and evaluative conception of human purposiveness. Mulcaster appreciated the mechanical, means-oriented, capacity in specific educational programs. Education could, if it was attuned to specific social conditions, facilitate each person's reaching his or *her* prescribed ends. Whatever a boy or a *girl* was to be, a well-defined education would help. But Mulcaster did not define any value in education-in-itself. (I emphasize the feminine context here because Mulcaster publicly favored female education.) See p. 53.

79. Ibid., p. 21.

80. Ibid., pp. 66 and 23.

81. Ibid., p. 66.

82. Ibid., p. 69.

83. Ibid., p. 41.

84. Ibid., p. 47.

85. If Lacey Baldwin Smith is correct, no more than did the *Homilies* did education accomplish its ideological task. He claims in *Treason in Tudor England* that possibly "at no time in human history has the gulf between the educational goals of society and the reality existing outside the classroom been so great" (p. 73). Mulcaster's educational theories, then, can be seen as part of the cultural process of meaning redefinition which this "gulf" but particularly evidences.

86. Ibid., p. 43.

87. Hilton's essays on "The English Virtuoso" broaden the perspective here. The "virtuoso" of the early seventeenth century pursued learning-in-itself for curiosity, delight, and pleasure. He looked back upon a sixteenth-century interest in learning, first dominated by Elyot and other humanists which emphasized gentlemanly moral and political ends to practical studies, and focused, during the Elizabethan years on courtly ornamental pursuits (Elyot superseded by Castiglione). In its development, then, the "English ideal of public

service" remained bound to learning throughout the Tudor years (p. 59). Baconian utilitarianism and much of the new science suggest the paths which modified virtuosi ideals as the century progressed. The development that Hilton charts—the close early Tudor relationship between learning and the public order, the self-conscious linking of learning to a rigidified, institutionalized order in Elizabethan England ("decorum"), learning as an avenue of individual expression in the early seventeenth century, and finally learning as the means to change or reform society as the century progresses—offers a fascinating and descriptive documentation for the related histories of self and order that I am tracing in this book.

88. Neale, *Elizabeth I and Her Parliaments*, p. 325.

89. Quoted in J. E. Neale, "The Commoners' Privilege of Free Speech in Parliament," in *Historical Studies of the English Parliaments*, eds. E. B. Fryde and E. Miller (Cambridge, Cambridge University Press, 1970), p. 168.

90. Quoted in Neale, *Elizabeth and her Parliaments*, p. 391.

91. Ibid., p. 321.

92. Ibid., p. 322.

93. Neale, "Free Speech in Parliament," p. 161.

94. Ibid., p. 171.

95. Ibid.

96. Ibid., p. 172.

97. Peter Wentworth, *A Pithie Exhortation onto Her Maiestie For Establishing Her Successor to The Crowne* (London, 1598), written early in 1590s, p. 8.

98. Ibid., p. 6.

99. Ibid., p. 5.

100. Ibid., p. 3.

101. Ibid., p. 1. Wentworth refers to "country" rather than to "commonweal" in reference to England. By doing this he continues to define the "public good" as the end of temporal action, but he succeeds in conflating the traditional concepts of "commonwealth" and "State" and thereby putting upon Parliament the legislative and thus the defining responsibility for directing man toward attaining this public good.

102. Ibid., p. 8.

103. Ibid., p. 27.

104. These are not offered here as absolutely antithetical propositions. Indeed, Puritan consciousness attacked wholesale the traditional idea of order, and as Walzer shows, much of this attack was socially and psychologically motivated. My point, however, is to emphasize the development of self-consciousness and the burden of responsibility for the secular world. These I reiterate, not Puritan theology, determined the social and the psychological structure of the "Secular State." See Walzer, *Revolution*, and my Introduction.

105. Wentworth, *Pithie Exhortation*, p. 26.

106. Ibid., p. 99.

107. Peter Wentworth, *A Treatise Containing M. Wentworth his judgment of the heire apparent* (London: 1598), written 1595, p. 53.

108. Ibid.

109. Ibid.

110. Ibid., pp. 53–54.

111. I discussed Greenleaf's work briefly in the Introduction; see also Chapter 4.

112. Wentworth, *Treatise*, pp. 49–50. Both Wentworth and Hobbes seem to reconcile "order" and social contract theory. Hobbes, however, derives his theories from an understanding of the psychological nature of individual relations, while Wentworth argues his from social expressions and relationships.

113. I will not attempt a full historiographical criticism here. My purpose in studying Hooker is to shed light on Elizabethan intellectual history and my general argument should make it clear that I understand this history to be a particularly dynamic and a conceptually innovative one. Consequently, I find interpretations of Hooker as a Thomist or as a "pre-Whig" misleading. Until very recently, however, such interpretations were the rule rather than the exception. Peter Munz, in *The Place of Hooker in the History of Thought* (London: Routledge and Paul, 1952), argues that Hooker was the "last great representative of the medieval natural law school" (p. 206). Hooker turned to medieval Thomism to refute the Augustinianism he saw in Puritanism, Munz believes (p. 49). Indeed, he maintains that during the sixteenth century, theorists revived the medieval confrontation of Augustinianism and Thomism: "There was hardly a single new argument—in the philosophical sense—on either side" (p. 57). Egil Grislis ties Hooker to Aquinas in "Richard Hooker's Image of Man," *Renaissance Papers* (Southeastern Renaissance Conference, 1963), pp. 73–84; and John S. Marshall presents Hooker as a "Renaissance Thomist" and as a "pure Aristotelian" in *Hooker and the Anglican Tradition: An Historical and Theological Study of Hooker's Ecclesiastical Polity* (Sewanee, Tenn.: University Press, 1963). The most formulated of the "pre-Whig" interpretations belongs to F. J. Shirley who sees Hooker as "the direct forerunner of Sidney, Locke and Rousseau" (p. 69), in *Richard Hooker and Contemporary Political Ideas* (London: 1949).

More recently Hooker has been viewed within the context of late sixteenth-century European ideas. Arthur Ferguson, while stressing Hooker's impact as a "transitional" theorist, offers a more "historical" interpretation in "The Historical Perspective of Richard Hooker: A Renaissance Paradox," *Journal of Medieval and Renaissance Studies* 3:1 (Spring 1973), pp. 17–49. More directly, W. D. J. Cargill Thompson attempts to refute scholars who see Hooker as a precursor of seventeenth-century ideas and to understand him in his own age. Thompson emphasizes the polemical nature of Hooker's works and ties him to his Anglican compatriots, John Jewel and John Whitgift, in "The Philosopher of the 'Politic Society': Richard Hooker as a Political Thinker," in *Studies in Richard Hooker*, ed. W. Speed Hill (Cleveland: Case Western Reserve University Press, 1972).

In *Richard Hooker and the Politics of a Christian England* (Berkeley: University of California Press, 1981), Robert K. Faulkner attempts to examine Hooker within his own historical context and without labeling him as a transitional figure that prefigured social contract theorists or as another Thomist or Aristotelian (p. 61). Faulkner's Hooker, however, emerges from this confused book as an "unconventional" Thomist/Aristotelian (p. 61), a "moderate Christian Aristotelian who is a "great alternative" to Bacon, Hobbes, and Locke (p. 8), a theorist cut off from recent contemporary opinion and influence whose work on politics is the product of only his "own opinions about what is best" (p. 164), and a dedicated theocrat "essentially medieval in tenor" (p. 138), who is (comparable to Christopher Hill's Archbishop Laud) attempting to resuscitate the Middle Ages (p. 145).

Without engaging any of these interpretations directly, I believe that the following analysis of Hooker's works will clarify how I disagree with their emphases.

114. Ferguson, "Hooker," pp. 17–18.

115. See the latter sections of this chapter for a discussion of "Catholic political philosophy" and the "Doleman" controversy.

116. Richard Hooker, *Of the Laws of Ecclesiastical Polity*, in *The Works*, 3 vols., ed. J. Keble 7th edition, revised by R. W. Church and F. Paget (Oxford, Clarendon Press, 1888), preface, 1:164

117. Hooker, *Laws*, Book I, 1:244.

118. Ibid., I, 1:246.

119. Ferguson, "Hooker," pp. 24–25.

120. Zeeveld, *Temper of Shakespeare's Thought*, pp. 32–33.

121. Hooker, *Laws*, Book I, 1:247.
122. Ibid., I, 1:248.
123. Ferguson, "Hooker," p. 36.
124. Hooker, *Laws*, Book III, 1:384.
125. Ibid., III, 1:385–387.
126. Ibid., III, 1:395–396.
127. Ibid., Book IV, 1:481.
128. Thompson, "Politic Society," p. 43.
129. Shirley, *Hooker and Contemporary Political Ideas*, p. 96. I do not necessarily endorse Shirley's claim for Hooker's originality here. It is not clear that he did "systematically" formulate such a basis. Indeed, it seems probable that Shirley is telescoping the intellectual history of this idea. Nonetheless, Hooker's discussions of the origins of society and social order and of legitimate authority offer enlightening insights into late sixteenth-century political consciousness.
130. Hooker, *Laws*, Book I, 1:222.
131. Ibid., I, 1:239.
132. Hooker, *Book VIII*, p. 170.
133. Ibid.
134. Hooker, *Laws*, Book I, 1:242.
135. Hooker seems to suggest that men were inclined both to sociableness and to self-interest and conflict. Their propensity to society appears to be "natural" or potential, a product of "Right Reason." Yet man must act to resist his "other" inclination in order to fulfill his "natural" character. Unlike Hobbes, Hooker is not satisfied with only efficient motivation for social order.
136. Hooker, *Laws*, Book I, 1:242.
137. Ibid.
138. Ibid.
139. Hooker, *Book VIII*, p. 185.
140. Ibid.
141. Ibid.
142. Ibid., p. 241.
143. Ibid., p. 160.
144. Ibid., p. 157.
145. Ibid., p. 171.
146. Ibid., p. 172.
147. Ibid., p. 175.
148. Ibid., p. 279.
149. Ibid., p. 183.
150. Ibid., p. 178.
151. Two of the more recent contributions to Hooker scholarship have belabored this theme. See H. C. Porter, "Hooker, the Tudor Constitution and the *Via Media*," in *Studies in Richard Hooker*, ed. Hill, and Gerald R. Cragg's chapter, "Hooker, Andrewes, and the School of Laud," in his *Freedom and Authority: A Study of English Thought in the Early Seventeenth Century* (Philadelphia: Westminster Press, 1975). This latter book, in fact, is a simplified rehashing of traditional opinions about the political thought of Elizabethan and Early Stuart England. Of some interest are the chapters on Catholic political thought, a subject often overlooked by Tudor-Stuart intellectual historians. Interestingly, the new literary historicist critics who are busily deconstructing the ideology of order, Shakespeare, and a spate of other literary figures, still off-handedly reference Hooker as a "traditionalist." See for example, Moretti, "Tragic Form," p. 11.
152. Hooker, *Book VIII*, p. 168.

153. Ibid.
154. Ibid.
155. Ferguson, "Hooker," p. 40.
156. Hooker, *Laws*, Book VII, 3:226.
157. Ibid., VII, 3:250–251.
158. Ibid., Book V, 2:207.
159. Ibid., V, 2:208.
160. See Chapter 2.
161. Hooker, *Laws*, Book V, 2:208–209.
162. Ibid., V, 2:210.
163. Hooker, *Book VIII*, p. 280.
164. Hooker, *Laws*, Book I, 1:254.
165. Ibid.
166. Ibid., I, 1:215.
167. Lovejoy's *The Great Chain of Being* is the best source for an intelligent discussion of the "Idea of Good" and "The Idea of Goodness," both of which characterized the divine order.
168. Hooker, *Laws*, Book I, 1:215, emphasis mine.
169. Ibid. From Lovejoy we know that "Goodness" described the variety and multiplicity of possibilities and potentialities in Creator and in Creation. Cf. also Marsilio Ficino, "Five Questions Concerning the Mind," in *The Renaissance Philosophy of Man*, eds. E. Cassirer, P. O. Kristeller, and J. H. Randall, Jr. (Chicago: University of Chicago Press, 1956), pp. 193–212.
170. Hooker, *Laws*, Book I, 1:215.
171. I am not suggesting any link between these three Elizabethans. It is interesting to realize, however, that each created a striving and aspiring character, but each placed his character at a separate perspective in relation to the received idea of order. Hooker was familiar with Hermetic literature, though, and freely quoted Hermes Trismegistus throughout the *Laws*.
172. Hooker, *Laws*, Book I, 1:200.
173. Ibid., I, 1:202.
174. Ibid.
175. Ibid., I, 1:255.
176. Ibid.
177. For a discussion of metaphor, discourse, and cultural meaning see White, *Tropics of Discourse: Essays in Cultural Criticism*, p. 47.
178. W. Speed Hill, "The Evolution of Hooker's Laws of Ecclesiastical Polity," in *Studies in Richard Hooker*, p. 153.
179. Hooker, *Laws*, Book II, 1:322.
180. See Hill in *Studies in Richard Hooker*, p. 152.
181. Hooker, *Laws*, Book I, 1:222.
182. Ibid., I, 1:221.
183. Ibid., I, 1:220.
184. Ibid., I, 1:291.
185. Ferguson, "Hooker," p. 47.
186. Ibid., p. 49. I single Ferguson out for criticism on this point because his is the most "historical" and perceptive discussion of Hooker. In his more recent work, Ferguson suggests a better fit for Hooker in Elizabethan ideas; see *Clio*, pp. 207–222.
187. Shirley, *Hooker and Contemporary Political Ideas*, pp. 69, 99.
188. Parsons (R. Doleman), *A Conference about the Next Succession to the Crown of England* (London, 1681), preface, (first printed, 1594).

189. John Neville Figgis, *The Divine Right of Kings* (Cambridge: Cambridge University Press, 1922), p. 104. In the 1703 edition of Sir Thomas Craig's anti-Doleman pamphlet (see note 190), the preface links Doleman to the anti-king forces in the 1640s and names Milton and Harrington as Doleman's successors!

190. Sir Thomas Craig, *The Right of Succession To the Kingdom of England: Against the Sophisms of Parsons the Jesuite* (*Doleman*). 1703 publication (first published in 1603). Sir J. Hayward, *An Answer to the First Part of a Certaine Conference Concerning Succession, Published not long since under the name of R. Doleman* (London, 1603). Of note is the frontispiece to this work. It is the same design found on the title page to Forset's 1606 major work (see Chapter 4), and closely associated with mystical, Hermetic, and possibly Rosicrucian pamphlets: that is, the Hebrew letters for Jehovah, which resist actual pronunciation, set in clouds. Frances Yates deals with this symbol in *The Rosicrucian Enlightenment*. See Chapter 4.

191. Parsons, *A Conference*, p. 169.

192. Ibid.

193. Ibid., p. 170.

194. Ibid., p. 171.

195. See Walzer, *Revolution*, on the psychology of Puritan political behavior.

196. Parsons, *A Conference*, p. 59.

197. Ibid., p. 1.

198. Ibid., pp. 3, 6.

199. Ibid., p. 12

200. Ibid., p. 14.

201. Ibid., p. 10.

202. Ibid., p. 18.

203. Ibid., p. 63.

204. Ibid., p. 67.

205. Ibid., p. 108.

206. Ibid., pp. 54, 28.

207. Ibid., p. 58.

208. Ibid., p. 31. Notice Doleman's total misunderstanding or, perhaps, misrepresentation of the formal correspondence analogy of head and body of the commonweal. For him the idea of order is the authority of one Truth, and it allows for only one concrete manifestation.

209. It has been noted that a proper regard for an understanding of human history characterized radical theories in the seventeenth century. These characteristics are included in Greenleaf's "empiricism"; see *Order, Empiricism and Politics*. Also see Christopher Hill, *The Intellectual Origins of the English Revolution* (London: Oxford University Press, 1965.

210. Allen, *History*, pp. 256–260.

211. Craig, *Right of Succession*, p. 6.

212. Ibid.

213. Ibid., p. 10.

214. Hayward, *An Answer*, Sig. B4.

215. Craig, *Right of Succession*, p. 10.

216. Ibid., p. 113.

217. Ibid., p. 122.

218. Ibid., p. 30.

219. Hayward, *An Answer*, Sig. B2.

220. Ibid., Sig. A3.

221. Craig, *Right of Succession*, p. 2.

222. Ibid., p. 4.

223. Ibid., p. 3.
224. Ibid., p. 2.
225. Ibid., p. 71.
226. Hayward, *An Answer*, Sig. E–F.
227. Craig, *Right of Succession*, p. 203.
228. Ibid., p. 200.
229. Hayward, *An Answer*, Sig. N2, O2.
230. Craig, *Right of Succession*, p. 185.
231. Ibid., p. 193.
232. Ibid., p. 78.
233. Gordon J. Schochet, in *Patriarchalism in Political Thought: The Authoritarian Family and Political Speculation and Attitudes Especially in Seventeenth-Century England* (New York: Basic Books, 1975), situates the Doleman controversy in patriarchal context. Doleman and Hayward "foreshadowed the writings of patriarchalists and their critics," he concludes (p. 47). Schochet's analysis reduced to this context offers nothing useful about this controversy or about Catholic political thought. For a workmanlike introduction to Catholic political thought which, however, does not engage the Doleman controversy, see Peter Holmes, *Resistance and Compromise: The Political Thought of the Elizabethan Catholics* (Cambridge: Cambridge University Press, 1982). J. P. Sommerville discusses Catholic political ideas in *Politics and Ideology in England, 1603–1640* (New York: Longman, 1986). My book was in print by the time I saw Sommerville's work.
234. Craig, *Right of Succession*, p. 179.
235. Ibid., p. 168.
236. Hayward, *An Answer*, Sig. K3.
237. Ibid., Sig. K4.
238. Ibid., Sig. O4.
239. Ibid., Sig. P.

Chapter 4

1. Allen, *A History of Political Thought*, p. 249.
2. The fullest discussion is still that by Figgis, *Divine Right of Kings*. For a controversial appraisal of divine right theories and theorists, see Greenleaf, *Order, Empiricism and Politics*. Cf. Sommerville, *Politics and Ideology*, ch. 1.
3. I will discuss Filmer's political theories in Chapter 5.
4. James I, "The Trew Law of Free Monarchies: Or The Reciprock and Mutuall Duetie Betwixt A Free King, and His Naturall Subjects," in *The Political Works of James I*, ed. C. H. McIlwain (Cambridge: Harvard University Press, 1918), p. 54. All references to James I's writings are from this edition of his works.
5. James I, "Speech to Parliament, 1609," p. 307.
6. Ibid.
7. Figgis, *Divine Right*, p. 4.
8. Ibid., p. 140.
9. Ibid., p. 141.
10. McIlwain, *Political Works of James I*, introduction, p. xxx.
11. James I, "Basilikon Doron," p. 11.
12. Ibid., p. 15.
13. James I, "Opening Speech to Parliament, 1603," p. 270.
14. See brief discussion of merchant propaganda concerning this issue later in this chapter.
15. James I, "Speech to Parliament, 1609," p. 318.

16. James I, "Basilikon Doron," p. 18.

17. Ibid.

18. This was an important issue in England during these years. Both Coke and Bacon addressed it in the Calvin Case which concerned the Post-nati. Legally there were two defining natures to kingship: the *natural* referred to the physical person, and the *political* referred to the institution. Coke decided that "our liegence is to our natural liege Sovereign, descended of the Blood Royal of the Kings of this Realm." See Coke, "Seventh Reports, Calvin's Case," vol. 4 in *The Reports of Sir Edward Coke*, ed. J. H. Thomas and J. F. Frazer (London, 1826).

The Calvin case was a test case for James I. It considered whether one born in Scotland after James' accession to the English crown would be English and could inherit English lands. Aliens could not. Since Parliament had dashed James' hopes of an English-Scottish union, James wanted yet to gain rights to English citizenship for Scots. For a good discussion of Coke in relation to this case, see Charles Gray's "Reason, Authority and Imagination: The Jurisprudence of Sir Edward Coke," in *Culture and Politics From Puritanism to the Enlightenment*, ed. Perez Zagorin (Berkeley, Calif.: University of California Press, 1980).

Bacon agreed with Coke, but he stressed the superiority of natural law over common law and statute law. He maintained that allegiance is to the person of the king, and it is greater than laws of kingdoms. Francis Bacon, *The Works of Francis Bacon*, ed. James Spedding, R. L. Ellis, and D. D. Heath (London: Longmans, 1857–1859), (hereafter cited as Bacon *L.P.W.* [*Literary and Professional Works*] and *P.W.* [*Philosophical Works*]). *L.P.W.*, 2:641–667.

In a curious deconstructionist reading, Jonathan Goldberg, *James I and the Politics of Literature* (Baltimore: Johns Hopkins University Press, 1983), maintains that in Renaissance England "private" did not mean "individual" or "self"; rather it indicated "unavailability," "invisibility." Private is the negation of existence—not seen, since there is always an audience. He argues that James' "individuality" is not marked by his being a private person or having a self, but by incorporating God's seeing characteristics while not being always seen in public, and "he also bears another body, his invisible body, the body of his power" (pp. 148–149).

19. James I, "Basilikon Doron," p. 18.

20. James I, "Speech to Parliament, 1609," p. 309. James' remarks about the private and public identity of the Pope are revealing here. He doubts those who argue that "as a man, the Pope may fall into error, but not as Pope." Why, then, does not the Pope reform the man? James asked. And he continued: "But whether a king be deposed by that man the Pope, or by that Pope, the man, is it not all one?" James seems able to admit that the person and the public figure are separately received but that really there is only one, the private person who acts and is acted upon. He is struggling, then, to reaffirm a public identity and purpose for his self-conscious self. In some ways James reminds one of Shakespeare's Richard II, but he is more compromising with his identity than is Richard. Perhaps this is because he knows himself better. James I, "A Remonstrance for the Right of Kings," p. 209.

21. James I, "Trew Law," p. 46.

22. Ibid., p. 61.

23. Ibid., p. 62.

24. James I, "Speech to Parliament, 1609," p. 309.

25. James I, "Trew Law," p. 63.

26. James I, "Speech to Parliament, 1609," p. 310. Goldberg, in *James I*, carries Pocock's ideas about the political nature of formal language to an extreme conclusion. His central topic is "the ideological function of writing as an instrument of royal power" (p. 55). Following Derrida, he claims that discourse is self-referential, that speech repre-

sents itself, and then, that it is "in language that the king represents himself; it is in language that power is displayed" (p. 153). He cites from a 1607 speech to Parliament to illustrate: "Here I sit and governe it [Scotland] with my Pen, I write and it is done, and by a Clearke of the Councell I governe Scotland now, which others could not doe by the sword" (p. 55).

27. Figgis, *Divine Right*, p. 154.

28. Ibid., p. 153.

29. Edward Forset, *A Comparative Discourse of the Bodies Nátural and Politique* (London, 1606), introduction. On the title page is an engraving with the Hebrew letters that represent the unpronouncable name of God emblazoned in clouds. This symbol was used by those involved in Hermetic or Rosicrucian concerns. Yates discusses its esoteric meaning and the secrecy surrounding it in her works, especially *The Rosicrucian Enlightenment*. It is curious to see it, and to find other obvious Hermetic allusions, in a work such as this. Yates makes no reference to Forset in any writings of which I am aware. Yet she has tried to find a relationship between James' court and the mysteries of the supposed Hermeticists and Rosicrucians. In any event, these allusions color all that Forset says, and demonstrate that radical theories were often closely related to traditional political concerns during these years.

30. Ibid., p. 1.

31. Ibid., introduction.

32. Ibid.

33. Ibid., p. 3. I have pointed before to the importance that a balanced active-passive quality in order had for many late Renaissance neoplatonic and Hermetic theorists. Shakespeare's *Tempest* is structured along these lines. See Chapter 2. Bacon emphasizes this quality in *New Atlantis* and quotes the renowned Renaissance magician Cornelius Agrippa on this relationship. See the latter part of this chapter.

34. Forset, *Discourse*, p. 38.

35. Ibid., p. 39.

36. The analogy to alchemy, here, should not be overlooked. As Yates and others have shown, the operative qualities of the Renaissance alchemists and magicians helped facilitate the acceptance of man as the organizer and the definer in the new science as well as in the political world. Bacon certainly fits, here. Again, Prospero should be remembered.

37. Forset, *Discourse*, p. 41. Forset used natural analogy to strengthen the concept of cycles which had been revived by Renaissance classicists such as Machiavelli. In so doing, he tied the purpose of civil society ever more tightly to temporal definition, but restricted it to an organic cycle. Although this divorced civil order from any transcendental order it prevented a full appreciation and understanding of the purposive possibilities in civil order. Raleigh, too, was haunted by this world of ineluctable decay. Social order, then, remained a measure to resist inevitable decline.

38. Ibid., p. 48.

39. Bacon, "Advancement of Learning," in *P. W.*, 3:430.

40. Bacon, in *L & L*, 3:77.

41. Bacon, "Advancement of Learning," in *P. W.*, 3:441.

42. Bacon, in *L & L*, 3:26.

43. Ibid., p. 27.

44. S. R. Gardiner, ed. *Parliamentary Debates in 1610* (London: Camden Society, 1862), p. 24. I am more interested here in showing how the traditional idea of society and commonplace concepts were being publicly redefined than in detailing the specific history of this early process of redefinition. Of interest, however, is a comparative look at Cowell's 1607 edition of *The Interpreter*, and a 1708 edition titled *A Law Dictionary or, The*

Interpreter. The 1708 edition changed all of Cowell's references to the king being above the law. Under *King*, the 1708 edition stated: "And though for the better and more equal course in making Laws, He do admit the Three estates . . . yet this derogate not from his Power; for whatever they Act, He by his negative Voice may quath." Cowell's definition did not restrict the king's power to his veto, but rather, expressed James' opinion concerning the king's relation to Law: "Hee is above the Law by his absolute power . . . and though for the better and equall course in making Lawes, hee doe admit the 3 Estates . . . yet this, . . . is not of constraint, but of his own benignitie. . . . For otherwise were he a subject after a sort and subordinate, which may not be thought without breech of dutie and loyaltie. For then must we deny him to bee above the Lawe." Other comparisons are equally enlightening. See *Parliament* and *Subsidy*, for examples. For a fuller discussion, see Sommerville, *Politics and Ideology*, pp. 121–127.

45. Bacon, in *L & L*, 4:177.

46. Ibid., p. 280.

47. Ibid., p. 366.

48. Raleigh, *Works*, p. 19.

49. Bacon, in *L & L*, 4:371–372.

50. There is a parallel between changes in political philosophy and in natural philosophy during these years. I am not suggesting any causal relationship, though. Hobbes found that Galileo's ideas fitted what he understood about the laws of political nature.

51. Raleigh, *Works*, p. 59.

52. "Speech of Sir Henry Marten to Parliament, May 22, 1628" (Printed copy in British Museum), p. 5.

53. Henry Wright, *The First Part of The Disquisition of Truth, Concerning Political Affairs* (London, 1616), p. 12.

54. Ibid., pp. 13–14.

55. Raleigh, *Works*, p. 93.

56. With the breakdown of the traditional idea of order man might find himself thrust into such a role of self-responsibility. Marlowe's characters try to assume such a burden. The concept of cycles offered an alternative to this as well as to traditional ideas. Theorists conceived of various relationships between man and fortune. Francesco Guicciardini's *The History of Italy*, trans. and ed. S. Alexander (New York: Macmillan, 1969), stands out as a work in which Fortune appears totally undirected and nonpatterned. History offers no laws, no cycles. Man may act but Fortune acts despite man. Man's possibilities, then, remain heroically tragic because virtue and honor cannot be derived from practical considerations and actions. Machiavelli, in his *History of Florence* (New York: Harper, 1960), believed in historical laws and political laws. Man became the principle actor who could engage Fortune and affect the outcome of history. For an interesting interpretation of the idea of cyclical history in Tudor England, see Ferguson, *Clio*, ch. 10.

57. Raleigh, *Works*, p. 93.

58. Ibid., p. 138.

59. Bacon, "Of Kingdoms and Estates," in *L.P.W.*, 1:452.

60. Bacon, *L & L*, 3:105.

61. Ibid.

62. Ibid., p. 107.

63. Bacon, "Of Innovations," in *L.P.W.*, 1:433.

64. Ibid.

65. Bacon, "Of Vicissitudes of Things," in *L.P.W.*, 1:512.

66. Ibid., "Of Dispatch," in *L.P.W.*, 1:434.

67. Bacon, "Preface" to *De Sapientia Veterum*, in *L.P.W.*, 1:695.

68. Ibid., p. 646.

69. Ibid.
70. Bacon, "Styx; or Treaties," in *L.P.W.*, 1:707.
71. Bacon, "Of Truth," in *L.P.W.*, 1:378.
72. Bacon, "Of Sedition and Troubles," in *L.P.W.*, 1:409.
73. Wright, *Disquisition of Truth*, p. 87.
74. Fulbeck, *Pandectes*, pp. 68–69.
75. Bacon, "Of Sedition and Troubles," in *L.P.W.*, 1:410.
76. Bacon, "Advancement of Learning," in *P.W.*, 3:459.
77. Ibid., p. 460.
78. Ibid., p. 461.
79. Bacon, "Of Delays," in *L.P.W.*, 1:428.
80. Bacon, "Advancement of Learning," in *P.W.*, 3:466.
81. Ibid., p. 462.
82. D. Tuvil, *Essaies Politicke and Morall* (London, 1608), Fol. 9–12.
83. Ibid., Fol. 22.
84. Ibid., Fol. 25.
85. Ibid., Fol. 37.
86. Ibid., Fol. 115.
87. Ibid., Fol. 15–16.
88. Ibid.
89. Ibid., Fol. 17.
90. Ibid., Fol. 125.
91. Bacon, "Of The Colours of Good And Evil," in *L.P.W.*, 2:77.
92. Bacon, "Advancement of Learning," in *P.W.*, 3:468.
93. Sir Walter Raleigh, "Instructions to His Son and to Posterity," in *Advice to a Son*, p. 20.
94. Bacon, "In Felicem Memoriam Elizabethae," in *L.P.W.*, 1:309.
95. Bacon, "Advancement of Learning," in *P.W.*, 3:469.
96. Bacon, "Of Great Place," in *L.P.W.*, 1:398.
97. Ibid., p. 399.
98. Bacon, "Of Wisdom For a Man's Self," in *L.P.W.*, 1:432. Compare this with Polonius' advice to Laertes in *Hamlet* I. iii. 78–80: "This above all, to thine own self be true/And it must follow as the night the day/Thou canst not then be false to any man."
99. Bacon, in *L.P.W.*, 1:433.
100. Sir John Eliot, *De Jure Majestatis or Political Treatise of Government*, written about 1628–1630, ed. Rev. Alexander B. Grosart (London, 1882). In *Politics and Ideology*, Sommerville notes that this tract summarizes a Latin work by Arnisaeus. Since the arguments in *De Jure* do not appear to coincide with those in *The Monarchy of Man*, Sommerville concludes that it probably does not represent Eliot's views (p. 158). Such a requirement of consistency is unnecessary. Besides, a reading of the two works as I offer here indicates a carefully formulated position.
101. Ibid., p. 31.
102. Ibid., p. 7.
103. Ibid., p. 3.
104. Ibid., p. 9.
105. Ibid.
106. Ibid., p. 10.
107. Ibid., p. 11.
108. Ibid., p. 15.
109. Ibid., p. 19.
110. Ibid.

111. Ibid., p. 20.
112. Ibid., p. 165.
113. Ibid., p. 175.
114. Sir John Eliot, *The Monarchy of Man*, ed. A. B. Grosart (London, 1879), c. 1632. Eliot was dying while he wrote this pamphlet and that may account, in part, for its somber, contemplative tone.
115. Ibid., p. 10.
116. Ibid., p. 12.
117. Ibid., p. 47.
118. Ibid., p. 42.
119. Ibid., pp. 48, 51.
120. Ibid., p. 56.
121. Ibid.
122. Ibid., p. 63.
123. Ibid., pp. 72–73, 76.
124. Ibid., p. 76.
125. Ibid., p. 105.
126. Ibid., pp. 106–108. These corruptions manifested themselves as fear, hope, joy, and sorrow, and they thwarted man's peace. Eliot did not acknowledge the motivating power of natural appetites or passions as did Hobbes. But his ethical psychology defined a state of peace marked by the conciliation of actives with passives, by affection and progress without motion. In this sense, in any event, he shared with Shakespeare and Bacon a burgeoning consciousness of necessary change.
127. Ibid., p. 108. Eliot argues that through the state man can oppose these corruptions with four pillars of strength: Justice, Prudence, Temperance, and Fortitude.
128. Ibid., p. 130. "Sorrow" may be understood as introspectively induced tragedy. Eliot, as did Shakespeare, noticed the human mania to fix things immutable.
129. Ibid., p. 134.
130. Eliot's implication that peace is "not-war" is Hobbesian.
131. Eliot, *Monarchy*, p. 135.
132. Ibid., p. 136.
133. Ibid., p. 188.
134. Ibid., p. 207.
135. Ibid., p. 222.
136. Ibid., p. 223.
137. Ibid., p. 224.
138. Ibid., p. 225.
139. Ibid.
140. Ibid., p. 226.
141. Joyce Oldham Appleby, *Economic Thought and Ideology in Seventeenth-Century England* (Princeton: Princeton University Press, 1978), pp. 18–19.
142. W. K. Jordan, *Philanthropy in England, 1480–1660* (London: George Allen & Unwin, 1960), pp. 143, 146. Jordan has been severely attacked for failing to compensate in his arguments for the inflation. A telling critique is William G. Bittle and R. Todd Lane, "Inflation and Philanthropy in England: A Re-Assessment of W. K. Jordan's Data," *Economic History Review*, 2nd ser. 29 (May 1976), pp. 203–210. Nonetheless Jordan's generalized account of merchant values is still useful.
143. Appleby, *Economic Thought*, p. 26.
144. Ibid., p. 4. An early seventeenth-century poetical defense of the merchant is found in the introduction to J. Browne's *The Merchants Avizo* (London, 1607):

When Merchants trade proceeds in peace,
And labours prosper well:
Then Common-weales in wealth increase,
As now good proofe can tell.

Let no man then grudge Merchants state,
Nor wish him any ill:
But pray to God our King to save,
And Merchants state helpe still.

Also, an interesting pamphlet that defends the status of city merchants and especially of apprentices is *The Cities Advocate: Whether Apprenticeship Extinguisheth Gentry?* (London, 1629), by E. Bolton. Bolton argues against Erasmus and Sir Thomas Smith, among others, who claim that gentlemen derogate their status when they enter commerce.

145. J. U. Nef, *Cultural Foundations of Industrial Civilization* (Cambridge: Cambridge University Press, 1958), p. 17.

146. See Sir Thomas Smith, *A Discourse of the Commonweal of This Realm of England.*

147. G. deMalynes, *The Center of the Circle of Commerce* (London, 1624), p. 5.

148. Nef, *Cultural Foundations*, p. 59.

149. Charles Wilson, *England's Apprenticeship, 1603–1763* (New York: St. Martin's Press, 1965), p. 12.

150. R. H. Tawney, in his introduction to Thomas Wilson, *A Discourse Upon Usury* (New York: A. M. Kelley, 1963), p. 121.

151. See Sylvia Thrupp, *The Merchant Class of Medieval London* (Ann Arbor: University of Michigan Press, 1962).

152. Ibid., p. 23.

153. Ibid., p. 26.

154. Ibid., p. 28.

155. Ibid., p. 259.

156. Ibid., p. 315.

157. Quoted in Lynn R. Muchmore, "An Analysis of English Mercantile Literature, 1600–1642," Ph.D. Dissertation, University of Wisconsin, 1968, p. 12.

158. See Mary Dewar's introduction to Smith's *Discourse*. See also A. B. Ferguson, "The Tudor Commonwealth and the Sense of Change," *Journal of British Studies* 3:1 (1963), pp. 11–35.

159. Smith, *Discourse*, pp. 33, 46, 80.

160. Wilson, *A Discourse Upon Usury*, p. 209.

161. Ibid., p. 178.

162. Quoted in R. H. Tawney, *Business and Politics Under James I* (Cambridge: Cambridge University Press, 1958), p. 80.

163. James I. "Basilikon Doron," p. 26.

164. Quoted in Tawney, *Business and Politics*, p. 79.

165. Ibid., p. 278.

166. Jordan, *Philanthropy*, p. 327.

167. Muchmore, *Mercantile Literature*, p. 284. Also, for an analysis of the relation between merchant literature and public policy in the seventeenth century, see the chapter on mercantile literature in Gunn, *Politics and the Public Interest in the Seventeenth Century*. Although his primary thesis differs from mine, Gunn's discussion of merchant literature corroborates my analyses in certain instances.

168. E. Misselden, *The Circle of Commerce* (London, 1623), dedication.

169. Ibid.

170. Ibid., p. 2.

171. E. Misselden, *Free Trade, or Means to Make Trade Flourish* (London, 1622), p. 21.

172. Ibid.

173. Ibid., p. 4.

174. Malynes, *Center*, p. 45.

175. Misselden, *The Circle*, pp. 17–18.

176. Ibid., pp. 16–17, 19–20. "What Misselden's arguments threatened was the accepted dividing line between the natural and the social." Appleby, *Economic Thought*, p. 47.

177. Ibid., p. 42.

178. Ibid., p. 19.

179. Smith, *A Discourse*, p. 20.

180. Wilson, *A Discourse Upon Usury*, p. 250.

181. John Wheeler, *A Treatise of Commerce* (London, 1601), p. 2.

182. Ibid., p. 4.

183. Ibid., p. 5.

184. G. deMalynes, *The Antient Law-Merchant* (London, 1629), dedication.

185. Ibid., Sig. A4.

186. Thomas Mun, *England's Treasure by Forraign Trade* (New York: Macmillan, 1895; facsimile of 1664 ed.), p. 3.

187. Ibid., p. 5.

188. Quoted in Jordan, *Men of Substance*, p. 220.

189. Quoted in ibid., p. 219.

190. Malynes, *The Antient-Law Merchant*.

191. Mun, *England's Treasure*, pp. 9–18.

192. Lewes Roberts, *The Treasure of Traffike*, 1641, in *Early English Tracts on Commerce*, ed. J. R. McCulloch (Cambridge: Cambridge University Press, 1954), p. 25.

193. Ibid., pp. 23–40.

194. Mun, *England's Treasure*, p. 7.

195. Roberts, *Treasure*, p. 29.

196. Wheeler, *Commerce*, p. 5.

197. Wilson, *A Discourse Upon Usury*, p. 203.

198. Roberts, *Treasure*, p. 7.

199. Ibid., p. 55.

200. G. deMalynes, *A Treatise of the Canker of England's Commonwealth* (London, 1601), p. 95.

201. D. Digges, *The Defence of Trade* (London, 1615), p. 2.

202. Fulbeck, *Pandectes*, p. 71.

203. Bacon, "Of Empire," in *L.P.W.*, 1:422. For similar comments see Tobias Gentlemen, *England's Way to Win Wealth, and to employ Ships and Mariners . . .* (London, 1614), p. 8. And Browne, *Merchants Avizo*, and Henry Parker's comments in Jordan, *Men of Substance*, p. 214.

204. Mun, *England's Treasure*, p. 3.

205. Roberts, *Treasure*, p. 3.

206. Jordan, *Men of Substance*, p. 216.

207. Ibid., p. 215.

208. Ferguson, "Tudor Commonwealth," p. 32.

209. Malynes, *Canker*, p. 95.

210. A. F. Chalk, "Natural Law and the Rise of Economic Individualism in England," *Journal of Political Economy* 59 (1951): p. 333.

211. Ibid., p. 338.
212. R. H. Tawney, in the introduction to Wilson's *A Discourse Upon Usury*, p. 110.
213. Wilson, ibid., p. 200.
214. Ibid., p. 248.
215. Ibid.
216. Ibid.
217. Ibid., p. 250.
218. Ibid., p. 235.
219. Mason, *A Mirrour For Merchants*, pp. 79–80.
220. Agnew, *Worlds Apart*, p. 121.
221. Ibid., pp. 121–122.
222. Miles Mosse, *The Arraignment and Conviction of Usurie* (London, 1595), p. 81.
223. Ibid., p. 102.
224. Ibid.
225. Mason, *A Mirrour For Merchants*, introduction.
226. Bacon's works and Shakespeare's plays are full of commercial language and images. The early scenes in *Troilus and Cressida* perhaps best exemplify this, but there are many other instances. For two "materialist" renderings of Shakespeare from this perspective, see Eagleton, *William Shakespeare*, ch. 5, and Dollimore, *Radical Tragedy*, pp. 195–203 (on *Lear*).
227. Mosse, *Arraignment*, p. 29.
228. Thomas Culpeper, *A Tract Against Usurie* (London, 1621), p. 10.
229. Mun, *England's Treasure*, p. 79.
230. Bacon, "Of Usury," in *L.P.W.*, 1:474.
231. Ibid., p. 475.
232. Ibid., p. 474.
233. Ibid., p. 475.
234. Ibid., p. 476.
235. Robert Filmer, *A Discourse Whether it may be Lawful to take Use for Money* (London, 1678), Preface, Sig. B5.
236. Ibid.
237. Ibid.
238. Ibid.
239. Ibid.
240. Ibid., p. 80.
241. Ibid., pp. 82–83.
242. I will deal at some length with Filmer's political thought in Chapter 5.
243. See Wheeler, *Commerce*.
244. Roberts, *Treasure*, p. 3.
245. John Roberts, *Trades Increase* (London, 1615), p. 51.
246. Mun, *England's Treasure*, p. 3.
247. Appleby, *Economic Thought*, p. 41.
248. Ibid., p. 14.
249. I have only mentioned thinkers already discussed in this book. Some of them are not commonly associated with this tradition. For a more detailed explanation of my reasons for including this discussion, see the Introduction.
250. John G. Burke, "Hermetism as a Renaissance World View," in *The Darker Vision of the Renaissance*, ed. Robert S. Kinsman (Los Angeles: University of California Press, 1974), especially pp. 96–102.
251. Ibid., p. 112.

252. P. J. French, *John Dee: The World of an Elizabethan Magus* (London: Routledge and Kegan Paul, 1972), p. 19.

253. Ibid., pp. 6, 30.

254. Ibid., p. 69. Although these tracts were extremely religious, even mystical in nature, a relationship should be considered between this gnostic philosophy and Voegelin's understanding of the essentially secular nature of just this type of gnostic aspiration. See my Introduction.

255. Ibid., pp. 75, 90.

256. Ibid., p. 161. French echoes Yates in maintaining that Hermetic science opposed Aristotelian science. Hermetic theory was operation-oriented; Aristotelian science was uninterested in practicality.

257. Ibid., p. 163.

258. Yates, *Giordano Bruno*, p. 144.

259. Ibid., pp. 155–156.

260. See ibid., p. 178; See also French, *Dee*, p. 155.

261. Yates, *Giordano Bruno*, p. 233. For a full elaboration of the Rosicrucian tradition in this context, see Yates, *The Rosicrucian Enlightenment*.

262. See discussion later in this chapter.

263. Yates, *The Rosicrucian Enlightenment*, p. 233.

264. I intend merely to mention a few well-known examples here and to catalogue a few less-known writers obviously influenced by this tradition. My object is to reveal the general and common interest in "Hermetic" themes and literature. See Chapters 2–4 for discussions that place Hooker, Shakespeare, Forset, and Sir John Eliot in this light.

265. French, *Dee*, pp. 137–138.

266. Ibid.

267. Ibid., p. 146.

268. Sir Philip Sidney, *An Apology for Poetry or the Defence of Posey*, ed. Geoffrey Shepherd (London: Thomas Nelson and Sons, 1965), p. 96.

269. French, *Dee*, p. 146. Cf. Levao's account of Sidney's "poesie" as invention and of Sidney as "self-consciously" privileging fiction in *Renaissance Minds*.

270. In the introduction to Sidney, *Apology*, p. 25.

271. Sir Fulke Greville, *The Life of the Renowned Sir Philip Sidney*, ed. Nowell Smith (Oxford: Oxford University Press, 1907), p. 44.

272. Ibid., p. 9.

273. Ibid., p. 18.

274. Ibid., p. 1.

275. Rees, *Fulke Greville, Lord Brooke, 1554–1628*, p. 70.

276. Greville, *Life of Sidney*, pp. 178–179, 197.

277. Also see Frank Manley's introduction to the poem for a fine discussion of the three-part structure of the poem. Although Manley does not make any reference to Hermetic influences, one's appreciation of the poems is furthered by such a consideration. *John Donne: "The Anniversaries"* (1611), ed. Frank Manley (Baltimore: Johns Hopkins University Press, 1963).

278. Quoted in ibid., p. 42.

279. Quoted in Paolo Rossi, *Francis Bacon: From Magic to Science*, trans. Sacha Rabinovitch (Chicago: University of Chicago Press, 1968), p. 19, emphasis is mine.

280. Rossi, ibid., p. 17.

281. Ibid.

282. Quoted in ibid., p. 21, emphasis is mine.

283. Thomas Acheley, *The Massacre of Money* (London, 1602), Sig. F3.

284. See Yates, *The Rosicrucian Enlightenment*, and Yates, *Shakespeare's Last Plays*. For other approaches to this theme and to Jacobean masques, see Ferguson, *Clio*, Ch. 10; Dollimore, *Radical Tragedy*, pp. 26–28; Goldberg, *James I*, ch. 2; and Stephen Orgel, *The Illusion of Power: Political Theatre in the English Renaissance* (Berkeley: University of California Press, 1975).

285. William Shakespeare, *King Henry the Eighth*, ed. S. Schoenbaum, Signet Classic Shakespeare, gen. ed. Sylvan Barnet (New York: New American Library, 1967), V. v. 41–54.

286. Three very different pamphlets published in the late Elizabethan years indicate Hermetic attitudes. G. Ripley, *The Compound of Alchymy*, ed. Ralph Rabbards, Gentleman, (1591). Rabbards reissued this treatise about alchemy and "Mercuries three" which had originally been written during Edward IV's time. Rabbards dedicated the pamphlet to Queen Elizabeth; he quickly mentioned John Dee and the *Hermeticam Veneretur*, and offered his work as service to God, Queen, and Country. But it was to the other gentlemen of England that he purposively dedicated the work: "Yours in the furtherance of Science." Also see Robert Mason, *Reasons Monarchie* (1602). Mason relies upon Hermes Trismegistus from the *Pimander* to compare man to God and to show that man is "an abridgeman of God, and the worlde": pp. 7, 21, 75. And see Anthony Gibson, *A Woman's Woorth . . . written by one that hath heard much, seen much, but knowes a greate deale more* (1599). This is a translation of a French pamphlet (see Ruth Kelso, *Doctrine for the Lady of the Renaissance* [Urbana: University of Illinois Press, 1956]).

287. Bacon, "The Great Instauration," in *P. W.*, 4:21.

288. Ibid., p. 24.

289. Bacon, "Praise of Knowledge," *L & L*, 1:123.

290. Rossi, *Bacon*, p. 9. For another look at Bacon's place in the Renaissance tradition especially concerning the relationship between knowledge and science, see Michael Hathaway, "Bacon and 'Knowledge Broken': Limits for Scientific Method," *Journal of the History of Ideas* 39:2 (April–June 1978): pp. 183–197.

291. Bacon, "The Great Instauration," in *P. W.*, 4:7.

292. Bacon, "Letter to Burghley," *L & L*, 1:109.

293. Bacon, "The Great Instauration," in *P. W.*, 4:7.

294. Bacon, "Novum Organum," in *P. W.*, 4:39.

295. Bacon, "The Great Instauration," in *P. W.*, 4:7.

296. Bacon, ibid., p. 28. One senses, however, that as with Hobbes' "state of nature," Bacon's *New Atlantis* may have been an effort that resulted from a marriage of these supposedly inimical methodologies. Bacon was never comfortable with the implications in his understanding of knowledge. He suggested that knowledge and action were equated. As he argued in *De Sapientia Veterum*, "whoever has a thorough insight into the nature of man may shape his fortune almost as he will" ("Sphinx; or Science," in *De Sapientia Veterum*, in *L. P. W.* I:757); yet he downplayed the existential nature of creation and action. In "Metis; or Counsel," he used the ancient myth to prove that action must spring from the king's head as did Athena from Jupiter's head. Action then equaled creation. The image is of the king and counsel in wedlock; as he argued, the decision is as "shaped in the womb" ("Metis; or Counsel," in *De Sapientia Veterum*, in *L. P. W.* 1:762). For Bacon, then, the creative process, action, and definition, remained the venue of the intellectual, conceptualizing process. For an interesting and informative discussion of essentialism and existentialism which sheds light on the consciousness of early seventeenth-century England, see Northrop Frye, *Fools of Time*, especially pp. 3–13.

297. In Rossi, *Bacon*, p. 163. Again one should note the strong salvational temper in Bacon's quest for a unity of knowledge and nature. Redemption, no less, is at stake; though

Bacon is surely moving toward locating Heaven and Hell within temporal existence. Nonetheless, he does not fully accept Marlovian transcendency, and he translates "Faustian" aspiration into a disciplined organization of knowledge. As he warned in the "Novum Organum": "God forbid that we should give out a dream of our own imagination for a pattern of the world" (quoted in Rossi, p. 163). But isn't that all that he finally (*New Atlantis*) can do without abandoning his belief in order?

298. Bacon, "Advancement in Learning," in *P. W.*, 3:395. See also "Aphorisms," no. 45, *P. W.*, 4:55.

299. Bacon, *P. W.*, 3:370.

300. Howard B. White, *Peace Among The Willows: The Political Philosophy of Francis Bacon* (The Hague: Martinus Nijhoff, 1968), p. 127.

301. Sidney Warhaft, "Bacon and the Renaissance Ideal of Self Knowledge," *Personalist* 44:4 (1963): p. 469.

302. Ibid.

303. Bacon, "Aphorisms," no. 4, *P. W.*, 4:47.

304. Bacon, "Pan; or Nature," *L. P. W.*, 1:709. This understanding of the relationship of the One to the Many resembles Bacon's discussion of "two natures of good" in Book 7, ch. 1, of *The Advancement of Learning* (*P. W.*, 5:7). Both conceptualizations can be seen in the context of Lovejoy's discussion of "good" and "goodness" in his *The Great Chain of Being*. Bacon defines his two goods this way: "The one as everything is a total or substantive in itself, the other as it is a part or member of a greater body." Bacon prefers the second "good" because he understands it to be more utilitarian, more common-good-oriented.

305. Bacon, see "Cupid; or The Atom," *L. P. W.*, 1:729.

306. Bacon, "Pan," *L. P. W.*, 1:712.

307. Ibid., p. 713. Here Bacon in an imaginative way has joined the neoplatonic conceptions of variety and order with the Hermetic tradition of natural magic and has infused them with a Faustian aspiration to aggrandizement, manipulation, and transcendence.

308. See ibid., p. 714.

309. Bacon, "Orpheus; or Philosophy," *L. P. W.*, 1:722.

310. Ibid.

311. Ibid.

312. Ibid.

313. White, *Peace*, p. 105.

314. Bacon, *The New Atlantis*, *P. W.*, 3:134. In the "Preface" to *De Sapientia Veterum*, Bacon claimed that fables are a middle ground between "hidden depths of antiquity and the days of tradition and evidence that followed . . . beneath no small number of the fables of the ancient poets there lay from the very beginning a mystery and an allegory." *L. P. W.*, 1:695–696.

315. White, *Peace*, p. 112.

316. Bacon, *The New Atlantis*, *P. W.*, 3:130.

317. Yates, *The Rosicrucian Enlightenment*, pp. 125–129. White discusses Bruno's influence on Bacon in *Peace*, pp. 127–131.

318. Yates, *The Rosicrucian Enlightenment*, p. 128.

319. Bacon, *The New Atlantis*, *P. W.*, 3:140.

320. Ibid., p. 136.

321. Ibid., p. 164. The emphasis on secrecy in *The New Atlantis* is strange in respect to Bacon's predilection for openness and cooperation. Recourse to Rosicrucian concerns for social reform provides one explanation for this supposed shift in attitude. See also p. 146.

322. Ibid., p. 137.
323. Ibid., p. 138.
324. White, *Peace*, p. 142.
325. Bacon, *The New Atlantis*, *P.W.*, 3:137.
326. Ibid., p. 146.
327. Ibid., p. 156.
328. Ibid., p. 147.

Chapter 5

1. Sir Robert Filmer, "Observations Concerning the Originall of Government: Observations on Mr. Hobbes' *Leviathan*," in *Patriarcha and Other Political Works of Sir Robert Filmer*, ed. Peter Laslett (Oxford: Oxford University Press, 1949), p. 239. All references to Filmer in this chapter are from this collection.

2. Ibid., p. 242.

3. Sir Robert Filmer, *Patriarcha, A Defense of the Natural Power of Kings Against The Unnatural Liberty of The People*, p. 59. Richard Tuck argues that new evidence supports an earlier dating of *Patriarcha*. For the manuscripts that Laslett edited, Tuck suggests a first draft date between 1628 and 1631 and a revision date before 1642. He argues that the first half may have been completed by 1615 to contribute to the Oath of Allegiance debate! See "A New Date for Filmer's Patriarcha," *Historical Journal* 29:1 (1986): pp. 183–186.

4. Filmer's *Patriarcha* retreats from the burden of self-consciousness with which Shakespeare and Bacon struggled. Paternity, as natural social order, denies the depth psychological sense of the individual's responsibility for his world. Filmer avoids facing this burden even though he appreciates the challenge of innovation. In this sense, *Patriarcha* is very reminiscent of More's *Utopia*. For a full treatment of the development of the theory of patriarchalism in England, see Schochet, *Patriarchalism in Political Thought*. Schochet wisely situates patriarchal theory amidst discussions concerning the separation of social and political reality in the early seventeenth century. And in response to doctrines advocating such differentiation he claims that "patriarchal doctrine . . . was transformed from a vaguely articulated societal theory into an intentional political ideology" (p. 55). In spite of identifying its pragmatic origins as an articulated theory (an ideology, even), he continues to label patriarchal theory (and Filmer) as traditional and anachronistic, thus missing the point that in "becoming" an ideology of traditions, patriarchalism innovatively moved beyond traditional ideas and can better be appreciated as a response to tradition than as a response to other emergent "radical" theories.

5. See Laslett's introduction to ibid., p. 11, for comments about Filmer in this context.

6. Ibid., p. 32.

7. Ibid., p. 35.

8. Ibid., pp. 5, 30. Again, Schochet denies Filmer's radical impact by reducing patriarchal theory to anachronism. He argues that "patriarchal theory, to the extent that it presupposed the chain of being [by failing "to distinguish between political and other kinds of phenomena"] was a remnant of the middle ages, one of the remnants that was to be discarded soon after its nature was discovered." *Patriarchalism in Political Thought*, p. 86. Such an interpretation is ahistorical because it evaluates patriarchal theory in relation to later developments in political ideas. Beyond this, it ignores the interesting point that by discursively refashioning patriarchal attitudes into an ideology, the theory of patriarchy evidences a self-conscious act which structurally denies the traditional paradigm it ostensibly affirms.

9. Ibid., p. 28. Filmer owed much of his reliance on civil law to his political master, Jean Bodin. See ibid., pp., 17, 18. For Filmer on usury, see Chapter 4 of this book.

10. John Selden, *Table Talk*, ed. F. Pollock (London, 1927), p. 7. All citations are from this edition. Selden never took the covenant or subscribed the Engagement, yet was respected and well treated by the government after 1649. See Tuck, *Natural Rights*, p. 99.

11. Laslett, "Introduction," in *Works of Sir Robert Filmer*, pp. 12, 31.

12. Filmer, *Patriarcha*, pp. 53–56. See Chapter 3 for a discussion of Catholic political philosophy.

13. Laslett, "Introduction," p. 17.

14. Ibid., p. 12.

15. Ibid., p. 13.

16. Filmer, *Patriarcha*, p. 55.

17. Ibid., p. 62. For a different look at the relationship between patriarchy and politics in the context of the growth of individualism during this period, see Stone's *The Family, Sex and Marriage*, especially chs. 5 and 6. Also see Jonathan Goldberg, *James I and the Politics of Literature*, and Margaret W. Ferguson, Maureen Quilligan, and Nancy J. Vickers, eds., *Rewriting the Renaissance: The Discourses of Sexual Difference in Early Modern Europe* (Chicago: University of Chicago Press, 1986).

18. Schochet, of course, claims that for Filmer, God's will alone justified all political and social activity, and because of this he had no theory of change and thus cannot be seen in relation to later conservative historicist positions. See *Patriarchalism in Political Thought*, p. 157. On the other hand, Margaret A. Judson claims that because of his belief that God justified political activity Filmer, as did other nonsecular advocates of the Commonwealth government, supported the Engagement. See *From Tradition to Political Reality: A Study of the Ideas Set Forth in Support of the Commonwealth Government in England, 1649–1653* (Hamden, Conn.: Archon Books, 1980), pp. 53–55.

19. Filmer, *Patriarcha*, p. 63. We can read authoritative as "author" and creator in this context as well as just enforcer of law. The king is indeed the "author" of society.

20. See Filmer, "The Anarchy of A Limited or Mixed Monarchy or A Succinct Examination of the Fundamentals of Monarchy . . ." (1648), p. 304. See Laslett's "Introduction," p. 18. Filmer finds support for much of this basic theory in Bodin.

21. Filmer, "Anarchy," p. 277.

22. Filmer, *Patriarcha*, p. 93.

23. Ibid., p. 93, and Filmer, "Anarchy," p. 277.

24. Filmer, "Anarchy," p. 279.

25. Filmer, *Patriarcha*, p. 105.

26. Filmer, "The Freeholder's Grand Inquest Touching our Soveraigne Lord the King and his Parliament" (1647), p. 171. Note in reference to laws, Filmer argues that kings *make* law, and Parliament *chooses* law.

27. Ibid., p. 172.

28. Ibid., pp. 156–157.

29. Selden directly refuted the idea that the prerogative is equal to the king's will.

30. In his effort to identify a developing historical consciousness during these years, Ferguson offers a most intelligent discussion of Selden, which focuses on Selden's awareness of historical relevance and change in regard to law, polity, and custom. He even suggests that Selden's methodology was pre-Burkean. *Clio*, pp. 57–59, 117–125, 292–298. So too does Tuck, *Natural Rights*, p. 84.

31. Filmer, *Patriarcha*, p. 110.

32. I have in mind here the widest meaning for property (propriety); as Locke argues: "By property I must be understood here, as in other places, to mean that property which men have in their persons as well as goods." John Locke, *The Second Treatise of*

Government, ed. Thomas P. Peardon (New York: Bobbs-Merrill, 1952), pp. 98–99. I also wish to suggest that property may be understood here in a psychoanalytical sense—as a term for an objectified self.

33. Norman O. Brown, *Love's Body* (New York: Vintage Books, 1966), p. 4.

34. Judson, *The Crisis of the Constitution*, pp. 386, 393.

35. Duddley Digges, *An Answer To A Printed Book Intitled Observations Upon Some of His Majesties late Answers and Expresses* (Oxford, 1642), p. 2. Tuck credits the "Tew Circle" of Oxford intellectuals (Seldenites) as joint authors of this tract. According to the copy he recognizes, along with Digges, Lord Falkland, and William Chillingworth share responsibility for it. Others credit Digges alone. See *Natural Rights*, p. 101. My intention here is briefly to focus upon some representative pamphleteers. Although many of these pamphlets respond directly to other pamphlets, and could be viewed within such a dramatic context. I organize them as Royalist, mixed, and parliamentary pamphlets for thematic purposes. Nevertheless, those I have chosen by Digges, Ferne, Hunton, Herle, and Parker are all dramatically and thematically interrelated and are clear and articulate examples of various points of view.

36. H. Ferne, *The Resolving of Conscience* (Cambridge, 1642), epistle.

37. Digges, *An Answer*, p. 46.

38. Ferne, *Resolving*, pp. 3, 23.

39. Ibid., p. 27.

40. Ferne, *Conscience Satisfied* (Oxford, 1643), p. 6.

41. Ibid.

42. Digges, *An Answer*, p. 51.

43. Ferne, *Resolving*, p. 17.

44. Ibid.

45. Ferne, *Conscience*, p. 8.

46. Ibid., pp. 10, 11.

47. Digges, *An Answer*, pp. 2, 6.

48. Ibid., p. 1.

49. Ibid.

50. In Judson, *The Crisis of the Constitution*, p. 395.

51. Ferne dramatized this view when he emphasized that the government, not the monarch, was mixed. See *Conscience*, p. 15. He acknowledged that the constitution recorded the change in government from absolute to mixed. And he reminded the Parliamentarians who wished to pretend that the present government, in fact, manifested an original contract and reflected an original constitution, that the constitution was not a precontrived embodiment of what presently existed (p. 11). But while Ferne recognized the transitional quality of the constitution and of the government (and many Parliamentarians refused to do this), he would not accept a meaning or purpose for government that was informed by this transitory nature.

52. Gunn, *Politics And the Public Interest*, p. 73.

53. Digges, *An Answer*, p. 6.

54. Parker argued the exact opposite. The king's interests depended on the public good. But here, too, the monarch and the people pursue goals that alienate each from the other. With both Digges and Parker there is a sense that nascent self-responsibility and the self-conscious act, as manifested in creative politics, necessitate alienation and initial isolation. I see these political theorists struggling with the burdens of self-interest and of personal, existential politics much in the way Cordelia and Coriolanus struggled with the burden of self-authenticity.

55. Digges, *An Answer*, pp. 10, 20.

56. Ferne, *Resolving*, p. 30.

57. Ibid., p. 32, and see also p. 12.

58. Digges, *An Answer*, p. 7.

59. Ibid., p. 1.

60. Judson, *The Crisis of the Constitution*, pp. 398, 405. Judson saw Hunton as the keenest analyst of English government as a mixed monarchy and of the crisis of this government, p. 397. Corrinne C. Weston supports this position in "The Theory of Mixed Monarchy under Charles I and After," *English Historical Review* 75 (1960): pp. 426–443. For a modification of this view, see later discussion.

61. Philip Hunton, *A Treatise of Monarchie* (London, 1643), p. 3. All citations from this edition.

62. Judson, *The Crisis of the Constitution*, p. 401.

63. Although Hunton seems to be describing a Hobbes-type contracted state here, fully repressed, with built-in inhibitors to revolt, he leaves himself an out. The contract assumes obligations between the people and the "Person" of the government—not the government. As he argues later, the people can resist the state, but not the person of the king.

64. Hunton was epistemologically unequipped to appreciate political reality as Hobbes did. In part I of the *Treatise* he described "Monarchies in Generall," and in part II he discussed "Of This Particular Monarchy." In decidedly non-Baconian fashion he explained that in this second part he would seek to define the "Composure" of "our Monarchy; for till we fully are resolved of that, we cannot apply the former generall Truths" (p. 30). So although he did not accept Filmer's or Ferne's depiction of the polity, he would use a similar methodology as they had in order to refute them.

65. Parker, *Observations*, p. 45.

66. Charles Herle, *A Fuller Answer to A Treatise Written by Doctor Ferne* (London, 1642), p. 2.

67. Judson, *The Crisis of the Constitution*, p. 423. For an introductory account of Parker's political views in relation to "order" and to human creativity, please see Chapter 1. Much of the remainder of this chapter bears a direct relationship to the arguments made there.

68. Ibid., p. 423.

69. Parker, *Observations*, p. 3.

70. Ibid., pp. 3, 5.

71. Herle, *A Fuller Answer*, p. 16.

72. Henry Parker, *The Case of Shipmony*, Speech to Parliament, November 3, 1640 (London, 1640), p. 2.

73. Ibid., p. 4.

74. Ibid., p. 5.

75. Ibid., p. 6.

76. Ibid., p. 9.

77. Parker, *Observations*, p. 2.

78. Parker, *Jus Populi*, p. 18.

79. Jordan, *Men of Substance*, p. 174.

80. Parker, *Jus Populi*, p. 18.

81. Judson, "Parker and Parliamentary Sovereignty," p. 142.

82. Parker, *True Grounds*, p. 8.

83. From *The Contra Replicant*, quoted in Judson, "Parker and Parliamentary Sovereignty," p. 157.

84. Quoted in ibid.

85. Ibid., p. 156.

86. Ibid., pp. 155–156.

87. Judson, *The Crisis of the Constitution*, pp. 409, 413. Judson has elaborated the idea that "a significant change took place in the way a number of Englishmen wrote about their polity." Specifically, she claims that theorists turned from constitutional to political thought to justify governments and in so doing established "secular approaches to human government" during the Civil War years. See *From Tradition to Political Reality*.

88. Herle, *A Fuller Answer*, p. 3.

89. Ibid., p. 4.

90. Ibid., p. 5.

91. Parker, *Observations*, p. 19.

92. Ibid. Until the idea of the state satisfactorily provided for a new positive self-society relationship, however, the picture of the state as an isolated institution prevailed. The society-state antithesis which Parker and Herle depicted manifested a still immature consciousness of the creative capacity of the self and of the secular sovereign state. Here the isolation of the state is analogous to the dramatic isolation of the self in Shakespeare's great tragedies from *Hamlet* to *Coriolanus*. As Prospero overcame this isolation (he conquered his island imagination) so too Hobbes reunited the individual and social nature in the sovereign state.

93. Parker, *Jus Populi*, pp. 3–4. See also Chapter 1.

94. Parker, *True Grounds*, p. 19.

95. Herle, *A Fuller Answer*, pp. 4, 8.

96. Ibid., p. 8.

97. Ibid., pp. 10–11, 15.

98. F. S. McNeilly, *The Anatomy of Leviathan* (New York: St. Martin's Press, 1968), p. 5.

99. Watkins, *Hobbes's System of Ideas*, pp. 84–85.

100. Hobbes, *Leviathan*, p. 84. All citations from Hobbes in this chapter are from *Leviathan*.

101. For a mythical (and a psychosexual) parallel of this creative process see Hesiod's *Theogony*, and Norman Brown's introduction to his translation of *Theogony* (New York: Bobbs-Merrill, 1953).

102. Macpherson, *The Political Theory of Possessive Individualism*, p. 78.

Conclusion

1. Julian Jaynes, *The Origins of Consciousness in the Breakdown of the Bicameral Mind* (Boston: Houghton Mifflin, 1976), p. 61.

2. Michel Foucault, *The Order of Things: An Archaeology of the Human Sciences* (New York: Random House, 1970), p. xx. In "Languages and their Implications," Pocock states that people "think by communicating language systems; these systems help constitute both their conceptual worlds and the authority structures, or social worlds, related to these; the conceptual and social worlds may each be seen as a context to the other, so the picture gains in concreteness. The individual's thinking may now be viewed as a social event, an act of communication and of response within a paradigm system, and as a historical event, a moment in a process of transformation of that system and of the interacting worlds which both system and act help to constitute and are constituted by" (p. 15). Here, I think, Pocock suggests the historically transformative character of thought that an intellectual history of consciousness and order can indicate.

3. White, "Foucault Decoded: Notes From Underground," p. 34.

4. Ibid.

5. White, "The Tasks of Intellectual History," p. 618.

6. Fred Weinstein and Gerald M. Platt, *Psychoanalytic Sociology* (Baltimore: Johns Hopkins University Press, 1973), p. 74.

7. Ibid., p. 85.

8. Brown, *Love's Body*, p. 112.

9. Voegelin, *The New Science*, p. 41.

10. Oakeshott, *Rationalism in Politics*, pp. 249–250. The following quoted passages are from these pages.

11. Michel Foucault, *The Order of Things*, p. 387.

12. Brown, *Love's Body*, p. 119.

BIBLIOGRAPHY OF WORKS CITED

Acheley, Thomas. *The Massacre of Money* (London, 1602).

Agnew, Jean-Christophe. *Worlds Apart: The Market and the Theater in Anglo-American Thought, 1550–1750* (Cambridge, 1986).

Allen, J. W. *A History of Political Thought in the Sixteenth Century* (London, 1941).

Anderson, Peter S. "Shakespeare's *Caesar*: The Language of Sacrifice," *Comparative Drama* 3 (1969).

Appleby, Joyce Oldham. *Economic Thought and Ideology in Seventeenth-Century England* (Princeton, 1978).

Aylmer, John. *An Harborowe for Faithful and Trew Servants* (London, 1559).

Bacon, Francis. *The Letters and the Life of Francis Bacon Including all his Occasional Works*, edited by James Spedding, 7 vols. (London, 1861–1874).

———. *The Works of Francis Bacon*, edited by James Spedding, R. L. Ellis, and D. D. Heath, 7 vols. (London, 1857–1859).

Baldwin, William. "Preface" to *Mirror for Magistrates*, edited by Lily B. Campbell (Cambridge, 1938).

Barber, C. L. "The Form of Faustus' Fortunes Good or Bad," in *Shakespeare's Contemporaries: Modern Studies in English Renaissance Drama*, edited by Max Bluestone and Norman Rabkin (Englewood Cliffs, N.J., 1970).

———. *Shakespeare's Festive Comedy: A Study of Dramatic Form and Its Relation to Social Custom* (Cleveland, Ohio, 1968).

Barbu, Zevedei. *Problems of Historical Psychology* (London, 1960).

Barckley, Sir Richard. *A Discourse of the Felicitie of Man* (London, 1598).

Baumer, F. Le Van. *The Early Tudor Theory of Kingship* (New Haven, 1940).

Bergeron, David Moore. *English Civic Pageantry, 1558–1642* (Columbia, S.C., 1971).

Bittle, William G., and R. Todd Lane. "Inflation and Philanthropy in England: A Re-Assessment of W. K. Jordan's Data," *Economic History Review* 2nd ser. 29 (May 1976).

Blake, William. *A Selection of Poems and Letters*, edited with introduction by J. Bronowski (Baltimore, 1958).

Bolton, E. *The Cities Advocate: Whether Apprenticeship Extinguisheth Gentry*? (London, 1629).

Breton, N. *Choice, Chance, and Change* (London, 1606).

Brown, Norman O. *Love's Body* (New York, 1966).

———. "XV. Kal. Mart. (February 15) LUPERCALIA," *New Literary History* 4:3 (Spring 1973).

Browne, J. *The Merchants Avizo* (London, 1607).

Burgess, Glenn. "Usurpation, Obligation and Obedience in the Thought of the Engagement Controversy," *Historical Journal* 29:3 (1986).

Burke, John G. "Hermetism as a Renaissance World View," in *The Darker Vision of the Renaissance*, edited with introduction by Robert S. Kinsman (Los Angeles, 1974).

Butterfield, Herbert. *The Origins of Modern Science, 1300–1800* (New York, 1957).
Caspari, Fritz. *Humanism and the Social Order in Tudor England* (Chicago, 1954).
Cavendish, George. *Metrical Visions* (London, 1825).
Cecil, William, Lord Burghley. "Certain Precepts for the Well Ordering of A Man's Life," in *Advice to a Son*, edited by L. B. Wright (Ithaca, N.Y., 1962).
Chalk, A. F. "Natural Law and the Rise of Economc Individualism in England." *Journal of Political Economy* 59 (1951).
Coke, Sir Edward. *The Reports of Sir Edward Coke*, edited by J. H. Thomas and J. F. Frazer, 6 vols. (London, 1826).
Collins, Stephen, L., and Alice L. Savage. "Consciousness, Order and Social Reality: Towards an 'Historical' Integration of the Human Sciences," *Humboldt Journal of Social Relations* 11:1 (1983–1984).
Collinson, Patrick. *The Elizabethan Puritan Movement* (London, 1967).
Cowell, John. *The Interpreter* (*Law Dictionary*) (London, 1637) (1607 edition).
———. *A Law Dictionary or, The Interpreter . . . much augmented and Improv'd . . .* (London, 1708).
Cragg, Gerald Robertson. *Freedom and Authority: A Study of English Thought in the Early Seventeenth Century* (Philadelphia, 1975).
Craig, Sir Thomas. *The Right of Succession To the Kingdom of England: Against the Sophisms of Parsons the Jesuite* (*Doleman*) (1603) (London, 1703).
Crompton, Richard, *The Mansion of Magnanimitie* (London, 1599).
———. *A Short declaration of the end of Traytors, and of false Conspirators against the State, & of the duetie of Subjects to theyr Sovereigne Governour . . .* (London, 1587).
Culpeper, Thomas. *A Tract Against Usurie* (London, 1621).
D'Ewes, Sir Simonds. *The Journals of all the Parliaments during the Reign of Queen Elizabeth, both of the House of Lords and House of Commons* (London, 1682).
Digges, D. *The Defence of Trade* (London, 1615).
Digges, Duddley. *An Answer To A Printed Book Intitled Observations Upon Some of His Majesties late Answers and Expresses* (Oxford, 1642).
Dollimore, Jonathan. *Radical Tragedy: Religion, Ideology and Power in the Drama of Shakespeare and His Contemporaries* (Chicago, 1984).
Donawerth, Jane. *Shakespeare and the Sixteenth-Century Study of Language* (Chicago, 1984).
Donne, John. *John Donne: "The Anniversaries,"* edited with introduction and commentary by Frank Manley (Baltimore, 1963).
Driscoll, James P. *Identity in Shakespearean Drama* (Lewisburg, Penn., 1983).
Eagleton, Terence. *Shakespeare and Society: Critical Studies in Shakespearean Drama* (London, 1970).
Eagleton, Terry. *William Shakespeare* (Oxford, 1986).
Eliot, Sir John. *De Jure Majestatis or Political Treatise of Government* (c. 1628–1630), edited by Rev. Alexander B. Grosart (London, 1882).
———. *The Monarchy of Man* (c. 1632), edited by A. B. Grosart (London, 1879).
Elton, G. R., ed. *The Tudor Constitution: Documents and Commentary* (Cambridge, 1968).
Elyot, Sir Thomas. *The Book named The Governor*, edited with introduction by S. E. Lehmberg (London, 1962).
Faulkner, Robert K. *Richard Hooker and the Politics of a Christian England* (Berkeley, Calif. 1981).
Febvre, Lucien. *The Problem of Unbelief in the Sixteenth Century: The Religion of Rabelais*, trans. Beatrice Gottlieb (Cambridge, Mass., 1982).

Ferguson, A. B. *Clio Unbound: Perception of the Social and Cultural Past in Renaissance England* (Durham, N.C., 1979).

———. "The Historical Perspective of Richard Hooker: A Renaissance Paradox," *Journal of Medieval and Renaissance Studies* 3:1 (Spring 1973).

———. "The Tudor Commonwealth and the Sense of Change," *Journal of British Studies* 3 (1963).

Ferguson, Margaret W., Maureen Quilligan, and Nancy J. Vickers, eds. *Rewriting the Renaissance: The Discourses of Sexual Difference in Early Modern Europe* (Chicago, 1986).

Fergusson, Francis. *The Idea of a Theater: A Study of Ten Plays, the Art of Drama in Changing Perspective* (Garden City, N.Y., 1953).

Ferne, H. *Conscience Satisfied* (Oxford, 1643).

———. *The Resolving of Conscience* (Cambridge, 1642).

Ferry, Anne. *The "Inward" Language: Sonnets of Wyatt, Sidney, Shakespeare, Donne* (Chicago, 1983).

Ficino, Marsilio. "Five Questions Concerning the Mind," in *The Renaissance Philosophy of Man*, edited by E. Cassirer, P. O. Kristeller, and J. H. Randall Jr. (Chicago, 1956).

Figgis, John Neville. *The Divine Right of Kings* (Cambridge, 1922).

Filmer, Robert. *A Discourse Whether it may be Lawful to take Use For Money* (London, 1678).

———. *Patriarcha and Other Political Works of Sir Robert Filmer*, edited with introduction by Peter Laslett (Oxford, 1949).

Floyd, Thomas. *The Picture of a perfit Common wealth* (London, 1600).

Forset, Edward. *A Comparative Discourse of the Bodies Natural and Politique* (London, 1606).

Foucault, Michel. *The Order of Things: An Archaeology of the Human Sciences* (New York, 1970).

French, Peter J. *John Dee: The World of an Elizabethan Magus* (London, 1972).

Fromm, Erich. *Escape from Freedom* (New York, 1941).

Frye, Northrop. *Fools of Time: Studies in Shakespearean Tragedy* (Toronto, 1967).

Fulbeck, William. *A Direction of Preparative to the Study of the Lawe* (London, 1600).

———. *The Pandectes of the Law of Nations* (London, 1602).

Fulbrook, Mary. *Piety and Politics: Religion and the Rise of Absolutism in England, Würtemberg and Prussia* (Cambridge, 1983).

Gardiner, S. R., ed. *Parliamentary Debates in 1610* (Camden Society, London, 1862).

Gay, Peter. *The Enlightenment: An Interpretation*, Vol. 1: *The Rise of Modern Paganism* (New York, 1966).

———. *The Enlightenment: An Interpretation*, Vol. 2: *The Science of Freedom* (New York, 1969).

Gentleman, Tobias. *England's Way to Win Wealth and to employ Ships and Mariners . . .* (London, 1614).

Gibson, Anthony. *A Womans Woorth . . . written by one that hath heard much, seen much, and knowes a greate deale more* (London, 1599).

Goldberg, Jonathan. *James I and the Politics of Literature* (Baltimore, 1983).

Gray, Charles. "Reason, Authority and Imagination: The Jurisprudence of Sir Edward Coke," in *Culture and Politics From Puritanism to the Enlightenment*, ed. Perez Zagorin (Berkeley, Calif., 1980).

Greenblatt, Stephen. "Invisible Bullets: Renaissance Authority and Its Subversion," *Glyph 8: Johns Hopkins Textual Studies* (Baltimore, 1981).

———. *Renaissance Self-Fashioning: From More to Shakespeare* (Chicago, 1980).

Greenleaf, W. H. *Order, Empiricism and Politics: Two Traditions of English Political Thought 1500–1700* (London, 1965).
Greville, Fulke. *The Life of the Renowned Sir Philip Sidney* (1652), edited and introduced by Nowell Smith (Oxford, 1907).
Grislis, Egil. "Richard Hooker's Image of Man," *Renaissance Papers* (Southeastern Renaissance Conference, 1963).
Guicciardini, Francesco. *The History of Italy*, translated, edited, with notes and introduction by Sidney Alexander (New York, 1969).
Gunn, J. A. W. *Politics and the Public Interest in the Seventeenth Century* (London, 1969).
Hanson, D. W. *From Kingdom to Commonwealth: The Development of Civic Consciousness in English Political Thought* (Cambridge, Mass., 1970).
Hathaway, Michael. "Bacon and 'Knowledge Broken': Limits for Scientific Method," *Journal of the History of Ideas* 39:2 (April–June 1978).
Hayward, Sir J. *An Answer to the First Part of a Certaine Conference Concerning Succession, Published not long since under the name of R. Doleman* (London, 1603).
Heller, Erich. "Faust's Damnation: The Morality of Knowledge," in *The Artist's Journey into the Interior and Other Essays* (New York, 1965).
———. "The Tempestuous Retreat," *The Listener* (January 21, 1965).
Herle, Charles. *A Fuller Answer to A Treatise Written by Docter Ferne* (London, 1642).
Hesiod, *Theogony*, translated, with an introduction, by Normon O. Brown (New York, 1953).
Hill, Christopher, *Intellectual Origins of the English Revolution* (Oxford, 1965).
Hill, W. Speed, "The Evolution of Hooker's Law of Ecclesiastical Polity," in *Studies in Richard Hooker*, edited by W. Speed Hill (Cleveland, 1972).
———, ed. *Studies in Richard Hooker* (Cleveland, 1972).
Hilton, Walter, E. Jr. "The English Virtuoso in the Seventeenth Century," *Journal of the History of Ideas* 3:1 & 2 (1942).
Hobbes, Thomas. *The Elements of Law Natural and Politic*, 2nd ed., edited by Ferdinand Tonnies, with new introduction by M. M. Goldsmith (London, 1969).
———. *English Works of Thomas Hobbes*, edited by Sir William Molesworth, 11 vols. (London, 1839–1845).
———. *Leviathan or the Matter, Forme and Power of a Commonwealth Ecclesiasticall and Civil*, edited with introduction by Michael Oakeshott (Oxford, 1946).
Holmes, Peter. *Resistance and Compromise: The Political Thought of the Elizabethan Catholics* (Cambridge, 1982).
Homilies (London, 1817).
Hooker, Richard. *Hooker's Ecclesiastical Polity, Book VIII*, Introduction by R. A. Houk (New York, 1931).
———. *The Works*, edited by J. Keble, 3 vols., 7th edition revised by R. W. Church and F. Paget (Oxford, 1888).
Hoopes, Robert. *Right Reason in the English Renaissance* (Cambridge, Mass., 1962).
Huizinga, Johan. *The Waning of the Middle Ages* (Garden City, N.Y., 1954).
Hunton, Philip. *A Treatise of Monarchie* (London, 1643).
James I. *The Political Works of James I*, introduced and edited by C. H. McIlwain (Cambridge, Mass., 1918).
Jaspers, Karl. *Tragedy Is Not Enough*, translated by Harold A. T. Reiche, Harry T. Moore, and Karl W. Deutsch (Boston, 1952).
Jaynes, Julian. *The Origin of Consciousness in the Breakdown of the Bicameral Mind* (Boston, 1976).
Jones, Whitney R. D. *The Tudor Commonwealth, 1529–1559* (London, 1970).

Jordan, W. K. *Men of Substance* (Chicago, 1942).
———. *Philanthropy in England, 1480–1660* (London, 1960).
Judson, M. A. *The Crisis of the Constitution* (New Brunswick, N.J., 1949).
———. *From Tradition to Political Reality: A Study of the Ideas Set Forth in Support of the Commonwealth Government in England, 1649–1653* (Hamden, Conn. 1980).
———. "Henry Parker and the Theory of Parliamentary Sovereignty," in *Essays in History and Political Theory* (Cambridge, Mass., 1936).
Kaufmann, R. J. "Ceremonies for Chaos: The Status of 'Troilus and Cressida,'" *English Literary History* 32:2 (1965).
Kaufmann, R. J., and Clifford J. Ronan. "Shakespeare's *Julius Caesar*: An Apollonian and Comparative Reading," *Comparative Drama* 4 (1970).
Kelso, Ruth, *Doctrine for the Lady of the Renaissance* (Urbana, Ill., 1956).
Kermode, Frank. "Introduction" to *The Tempest*, by William Shakespeare, 6th ed. *The Arden Shakespeare* (London, 1964).
Kramnick, Isaac. "Reflections on Revolution," *History and Theory* 11:1 (1972).
Kuhn, Thomas, S. *The Structure of Scientific Revolutions* (Chicago, 1970).
LaCapra, Dominick. *Rethinking Intellectual History: Texts, Contexts, Language* (Ithaca, N.Y., 1983).
Lambard, William. *Eirenarcha: or of The office of the Iustices of Peace in Foure Bookes* 2nd ed. (London, 1592).
Levao, Ronald. *Renaissance Minds and Their Fictions: Cusanus, Sidney, Shakespeare* (Berkeley, Calif., 1985).
Lever, J. W. *The Tragedy of State* (London, 1971).
Levin, Harry. *Christopher Marlowe: The Overreacher* (London, 1952).
Locke, John. *The Second Treatise of Government*, edited with introduction by Thomas P. Peardon (New York, 1952).
Lovejoy, Arthur. *The Great Chain of Being* (New York, 1960).
———. *Reflections on Human Nature* (Baltimore, 1969).
Lund, William, R. "Tragedy and Education in the State of Nature: Hobbes on Time and the Will," *Journal of the History of Ideas* 48:3 (1987).
McAlindon, T. *Shakespeare and Decorum* (New York, 1973).
Machiavelli, Niccolo. *History of Florence*, introduction by Felix Gilbert (New York, 1960).
McNeilly, F. S. *The Anatomy of Leviathan* (New York, (1968).
Macpherson, C. B. *The Political Theory of Possessive Individualism: Hobbes to Locke* (London, 1964).
Malynes, Gerard de. *The Antient Law-Merchant* (London, 1629).
———. *The Center of the Circle of Commerce* (London, 1624).
———. *A Treatise of the Canker of England's Commonwealth* (London, 1601).
Mandrou, Robert. *Introduction to Modern France, 1500–1640: An Essay in Historical Psychology*, translated by R. E. Hallmark (New York, 1977).
Marlowe, Christopher. *The Complete Plays of Christopher Marlowe*, edited with introduction and notes by Irving Ribner (New York, 1963).
———. *Plays and Poems*, edited and introduced by M. R. Ridley (London, 1973).
Marshall, John Sedberry. *Hooker and the Anglican Tradition: An Historical and Theological Study of Hooker's Ecclesiastical Polity* (Sewanee, Tenn., 1963).
Marten, Sir Henry. "Speech of Sir Henry Marten to Parliament, May 22, 1628."
Masinton, Charles G. *Christopher Marlowe's Tragic Vision* (Athens, Ohio, 1972).
Mason, Robert. *A Mirrour For Merchants—With an Exact Table to discover the excessive taking of Usurie* (London, 1609).
———. *Reasons Monarchie* (London, 1602).

Merbury, Charles. *A Briefe Discourse of Royall Monarchies, As of The Best Common Weale* (London, 1581).
Merleau-Ponty, Maurice. *Themes from the Lectures at the College de France 1952–1960*, translated by John O'Neil (Evanston, Ill., 1970).
Misselden, Edward. *The Circle of Commerce* (London, 1623).
———. *Free Trade, or Means to Make Trade Flourish* (London, 1622).
Montaigne, Michel de. *Essays*, translated with introduction by J. M. Cohen (Baltimore, 1958).
Montrose, Louis Adrian. "The Purpose of Playing: Reflections on a Shakespearean Anthropology," *Helios* n.s. 7:2 (1980).
More, Thomas. *Utopia*, translated with introduction by Paul Turner (London, 1965).
Moretti, Franco. "'A Huge Eclipse': Tragic Form and the Deconsecration of Sovereignty," *Genre* 15 (1982).
Morison, Richard. *A Remedy For Sedition* (London, 1536).
Morris, Christopher. *Political Thought in England: Tyndale to Hooker* (London, 1953).
Mosse, George L. *The Struggle for Sovereignty in England from the Reign of Queen Elizabeth to the Petition of Right* (New York, 1968).
Mosse, Miles. *The Arraignment and Conviction of Usurie* (London, 1595).
Muchmore, Lynn. "An Analysis of English Mercantile Literature, 1600–1642," Ph.D. Dissertation, University of Wisconsin, 1968.
Mulcaster, Richard. *The Educational Writings of Richard Mulcaster*, abridged and arranged by James Oliphant (Glasgow, 1903).
Mun, Thomas. *England's Treasure by Forraign Trade* (1664) (New York, 1895).
Munz, Peter. *The Place of Hooker in the History of Thought* (London, 1952).
Neale, Sir J. E. "The Commoners' Privilege of Free Speech In Parliament," in *Historical Studies of the English Parliament*, edited by E. B. Fryde and E. Miller (Cambridge, 1970).
———. *Elizabeth I and Her Parliaments 1559–1581*, (London, 1953).
Nef, J. U. *Cultural Foundations of Industrial Civilization* (Cambridge, 1958).
Norbrook, David. *Poetry and Politics in the English Renaissance* (London, 1984).
Norton, Thomas, and Thomas Sackville. *Gorboduc*, in *Medieval and Tudor Drama*, edited with introduction and modernizations by John Gassner (New York, 1963).
Oakeshott, Michael. *Rationalism in Politics* (London, 1962).
Orgel, Stephen. *The Illusion of Power: Political Theatre in the English Renaissance* (Berkeley, Calif., 1975).
Palmer, D. J. "Magic and Poetry in 'Doctor Faustus,'" *Critical Quarterly* 6:1 (1964).
Parker, Henry. *The Case of Shipmony*, Speech to Parliament, November 3, 1640 (London, 1640).
———. *Jus populi* (London, 1644).
———. *Observations upon some of his Majesties late Answers & Expresses*, 2nd ed. (London, 1642).
———. *The True Grounds of Ecclesiastical Regiment* (London, 1641).
Parsons (R. Doleman). *A Conference about the Next Succession to the Crown of England* (London, 1681).
Pitkin, Hanna Fenichel. *The Concept of Representation* (Berkeley, Calif., 1967).
Pocock, J. G. A. "England," in *National Consciousness, History and Political Culture in Early-Modern Europe*, edited by Orest Ranum (Baltimore, 1975).
———. *Politics, Language and Time: Essays on Political Thought and History* (New York, 1971).
Ponet, John. *A Short Treatise of politike power . . .* (Strasburg, 1556).
Porter, H. C. "Hooker, the Tudor Constitution and the *Via Media*," in *Studies in Richard Hooker*, edited by W. Speed Hill (Cleveland, 1972).

Pym, John. *A Speech delivered at a Conference of Lords & Commons—Jan. 25, 1641* (London, 1641).

Rabkin, Norman. *Shakespeare and the Common Understanding* (New York, 1967).

———. "Stumbling Toward Tragedy," in *Shakespeare's "Rough Magic": Renaissance Essays in Honor of C. L. Barber*, edited by Peter Erickson and Coppélia Kahn (Newark, N.J., 1985).

Raleigh, Sir Walter. "Instructions to His Son and to Posterity," in *Advice to a Son*, edited by L. B. Wright (Ithaca, N.Y., 1962).

———. *The Works of Sir Walter Raleigh, Political, Commercial, and Philosophical* (London, 1751).

Ratcliffe, Ægremont, trans. *Politique Discourses, treating of the differences and inequalities of Vocations, as well as Publique as Private* (London, 1578).

Rebholz, Ronald A. *The Life of Fulke Greville, First Lord Brooke* (Oxford, 1971).

Rees, Joan. *Fulke Greville, Lord Brooke, 1554–1628: A Critical Biography* (London, 1971).

Ribner, Irving. "Marlowe and Shakespeare," *Shakespeare Quarterly* 15:2 (1964).

Ripley, G. *The Compound of Alchymy*, edited by Ralph Rabbards (London, 1591).

Roberts, Lewes. *The Treasure of Traffike or A Discourse of Forraigne Trade*, in *Early English Tracts on Commerce*, edited by J. R. McCulloch (Cambridge, 1954).

Rossi, Paolo. *Francis Bacon: From Magic to Science*, translated by Sacha Rabinovitch (Chicago, 1968).

Sackville, Thomas. "Induction," in *Mirror for Magistrates*, edited by Lily B. Campbell (Cambridge, 1938).

Schochet, Gordon J. *Patriarchalism in Political Thought: The Authoritarian Family and Political Speculation and Attitudes Especially in Seventeenth-Century England* (New York, 1975).

Selden, John. *Table Talk*, edited by F. Pollock (London, 1927).

Shakespeare, William. *As You Like It*, edited by Albert Gilman, Signet Classic Shakespeare, General Editor, Sylvan Barnet (New York, 1963).

———. *Coriolanus*, edited by John Dover Wilson, *The New Shakespeare* (Cambridge, 1960).

———. *Hamlet*, edited by John Dover Wilson, *The New Shakespeare* (Cambridge, 1971).

———. *Henry IV Parts I and II*, edited by John Dover Wilson, *The New Shakespeare* (Cambridge, 1971).

———. *Julius Caesar*, edited by John Dover Wilson, *The New Shakespeare* (Cambridge, 1971).

———. *King Henry the Eighth*, edited by S. Schoenbaum, Signet Classic Shakespeare, General Editor, Sylvan Barnet (New York, 1967).

———. *King Lear*, edited by George Ian Duthie and John Dover Wilson, *The New Shakespeare* (Cambridge, 1960).

———. *Macbeth*, edited by John Dover Wilson, *The New Shakespeare* (Cambridge, 1947).

———. *A Midsummer Night's Dream*, edited by Wolfgang Clemen, Signet Classic Shakespeare, General Editor, Sylvan Barnet (New York, 1967).

———. *Richard II*, edited by John Dover Wilson, *The New Shakespeare* (Cambridge, 1971).

———. *The Tempest*, edited by Sir Arthur Quiller-Couch and John Dover Wilson, *The New Shakespeare* (Cambridge, 1971).

———. *Troilus and Cressida*, edited by Alice Walker, *The New Shakespeare* (Cambridge, 1957).

———. *The Winter's Tale*, edited by Sir Arthur Quiller-Couch and John Dover Wilson, *The New Shakespeare* (Cambridge, 1931).

Shirley, F. J. *Richard Hooker and Contemporary Political Ideas* (London, 1949).
Sidney, Sir Philip. *An Apology for Poetry or the Defence of Posey*, edited with introduction by Geoffrey Shepherd (London, 1965).
Skinner, Quentin. "Conquest and Consent: Thomas Hobbes and the Engagement Controversy," in *The Interregnum: The Quest for Settlement 1646–1660*, edited by G. E. Aylmer (Hamden, Conn., 1972).
Smith, Lacey Baldwin, *Elizabeth Tudor: Portrait of a Queen* (Boston, 1975).
———. *Henry VIII: The Mask of Royalty* (Boston, 1973).
———. *Treason in Tudor England: Politics and Paranoia* (Princeton, 1986).
Smith, Sir Thomas. *De Republica Anglorum*, edited by L. Alston (Facsimile of 1583 printing by Henrie Midleton for Gregorie Seton) (Cambridge, 1906).
———. *A Discourse of the Commonweal of This Realm of England* (1549), edited by Mary Dewar, Folger Shakespeare Library (Charlottesville, Va. 1969).
Sommerville, J. P. *Politics and Ideology in England, 1603–1640* (New York, 1986).
Starkey, Thomas. *England in the Reign of King Henry the Eighth: Dialogue*, edited by J. M. Cowper (London, 1878).
———. *Exhortation to the People Instructing Them to Unity and Obedience* (London, 1536).
Stone, Lawrence. *The Family, Sex and Marriage in England 1500–1800* (New York, 1977).
Strong, Roy. *The Cult of Elizabeth: Elizabethan Portraiture and Pageantry* (London, 1977).
———. *Splendor at Court: Renaissance Spectacle and the Theater of Power* (Boston, 1973).
Talbert, E. W. *The Problem of Order: Elizabethan Political Commonplaces and an Example of Shakespeare's Art* (Chapel Hill, N.C., 1962).
Tawney, R. H. *Business and Politics Under James I* (Cambridge, 1958).
Thomas, Keith. *Religion and the Decline of Magic* (New York, 1971).
Thompson, W. D. J. Cargill. "The Philosopher of the 'Politic Society': Richard Hooker as a Political Thinker," in *Studies in Richard Hooker*, edited by W. Speed Hill (Cleveland, 1972).
Thrupp, Sylvia L. *The Merchant Class of Medieval London* (Ann Arbor, Mich., 1962).
Tillyard, E. M. W. *The Elizabethan World Picture* (New York, n.d.).
Tuck, Richard. *Natural Rights Theories: Their Origin and Development* (Cambridge, 1979).
———. "A New Date for Filmer's Patriarcha," *Historical Journal* 29:1 (1986).
Tuvil, D. *Essaies Politicke and Morall* (London, 1608).
van den Berg, Kent T. *Playhouse and Cosmos: Shakespearean Theater as Metaphor* (Newark, N.J., 1985).
Voegelin, Eric. *The New Science of Politics* (Chicago, 1952).
———. *Order and History*, Vol. I: *Israel and Revelation* (Baton Rouge, La., 1956).
Waller, Garry. *English Poetry of the Sixteenth Century* (London, 1986).
Walzer, Michael. *The Revolution of the Saints: A Study in the Origins of Radical Politics* (New York, 1971).
Warhaft, Sidney. "Bacon and the Renaissance Ideal of Self Knowledge," *Personalist* 44:4 (1963).
Watkins, J. W. N. *Hobbes's System of Ideas* (London, 1965).
Weinstein, Fred, and Gerald M. Platt, *Psychoanalytic Sociology: An Essay on the Interpretation of Historical Data and the Phenomena of Collective Behavior* (Baltimore, 1973).
Wentworth, Peter. *A Pithie Exhortation onto Her Maiestie For Establishing Her Successor to The Crowne* (London, 1598).

———. *A Treatise Containing M. Wentworth his judgement of the heire apparant* (London, 1598).
Weston, Corrinne C. "The Theory of Mixed Monarchy under Charles I and After," *English Historical Review* 75 (1960).
Wheeler, John. *A Treatise of Commerce* (London, 1601).
Whetstone, George. *Aurelia. The Paragon of pleasure and Princely delights* (London, 1593).
———. *The English Myrror* (London, 1586).
———. *A Mirrour for Magistrates of Citties* (London, 1584).
White, Hayden. "Foucault Decoded: Notes from Underground," *History and Theory* 12:1 (1973).
———. "The Tasks of Intellectual History," *Monist* 53 (1969).
———. *Tropics of Discourse: Essays in Cultural Criticism* (Baltimore, 1978).
White, Howard B. *Peace Among the Willows: The Political Philosophy of Francis Bacon* (The Hague, 1968).
Williams, Robin M., Jr. "Change and Stability in Values and Value Systems," in *Stability and Social Change*, edited by Bernard Barber and Alex Inkeles (Boston, 1971).
Wilson, Charles. *England's Apprenticeship, 1603–1763* (New York, 1965).
Wilson, Thomas. *A Discourse Upon Usury*, edited and introduced by R. H. Tawney (New York, 1963).
———. *The Rule of Reason conteinying The arte of Logique* (London, 1551).
Wright, Henry. *The First Part of the Disquisition of Truth, Concerning Political Affairs* (London, 1616).
Yates, Frances A. *Giordano Bruno and the Hermetic Tradition* (London, 1964).
———. "Queen Elizabeth as Astraea," *Journal of Warburg and Courtauld Institutes* 10 (1947).
———. *The Rosicrucian Enlightenment* (London, 1972).
———. *Shakespeare's Last Plays: A New Approach* (London, 1975).
Zeeveld, W. G. *The Temper of Shakespeare's Thought* (New Haven, 1974).

INDEX

Abraham, 151
Acheley, Thomas, 142
Active/passive relationship. *See* Magic
Agnew, Jean-Christophe, 179, 183
Agrippa, Cornelius, 10, 139–142, 199
  Of the Vanitie and Uncertaintie of Artes and Sciences, 141
Alchemy, 139–141, 145. *See also* Hermeticism; Magic
Allen, J. W., 18, 106, 109
Allen, William, 94
Alston, L., 189
Alymer, John, 15–18, 26
Amphion, 140
Anderson, Peter S., 187
Appleby, Joyce, 127
Aristotle (Aristotelian), 86, 99, 130, 193. *See also* Thomism
Ascham, Anthony, 30
  The Bounds and Bonds of Public Obedience, 30
Astrology, 139. *See also* Hermeticism; Magic
Augustinianism, 193
Authority, idea of, 16, 30, 32–34, 75, 92–98, 103, 105, 107–109, 112–116, 123–125, 150, 152–158
  and Hooker, 95–98

Bacon, Francis, 12, 27, 40, 69, 74–75, 80, 83, 87, 109–110, 114–123, 125, 126–127, 134, 136, 138–148, 160, 168, 189, 192–193, 199, 202, 205, 207–209
  and Democritean cosmos, 119
  and existentialism, 80
  and fables, 119
  and Hobbes, 119, 122, 144, 146
  and knowledge, 143–145, 207
  and knowledge/order, 144–148
  and magic/science, 138, 146–148
  and nature/order, 145–148
  and self-interest, 80
  and Shakespeare, 121–123
  and time, 115, 117–119, 120
  and usury, 135–137
  works:
    "Cupid; or The Atom," 146
    De Sapientia Veterum, 119, 146, 207
    Essays, 120
    "Metis; or Counsel," 207
    New Atlantis, 126, 138, 141, 147–148, 189, 199, 207–208
      and knowledge, 147
      and magic, hermeticism, rosicrucianism, 147–148
      and nature, 147
      and social reform, 147–148
    Of The Colours of Good & Evil, 121
    "Of Great Place," 122
    "Of Wisdom For A Man's Self," 122
    "Orpheus; or Philosophy," 146
    "Pan; or Nature," 146
    Praise of Knowledge, 80
    "Sphinx; or Science," 207
    Styx, 119
Baldwin, William, 179
Barber, C. L., 45–47
Barbu, Zevedei, 171, 173
Barckley, Sir Richard, 83–86, 101
  A Discourse for the Felicitie of Man, 83
Baumer, F. Le Van, 176, 178
Bellarmine, Cardinal Robert, 151
Bergeron, David Moore, 178
Bittle, William G., 202
Blake, William, 183
Bodin, Jean, 210
Bohr, Niels, 187
Bolton, E., 203
Breton, Nicholas, 71
  Choice, Chance, and Change, 71
Brown, Norman O., 165, 183, 197, 213
  Love's Body, 165
Browne, J., 202
Bruno, Giordano, 11–12, 113, 140
Burgess, Glen, 181
Burke, John G., 205
Butterfield, Herbert, 175

Cabala, 140. *See also* Hermeticism; Magic
Calling, 111, 133

Calvin, John (Calvinism), 151, 161, 181
Calvin's Case, 198
Caspari, Fritz, 189
Catholic (literature, political philosophy), 92, 103, 151, 155. *See also* Doleman controversy
Cavendish, George, 26, 73
  Metrical Visions, 26
Cecil, Robert, 80, 116
Cecil, William, Lord Burleghy, 80, 144
Ceremony, 20, 81, 93, 105, 128
Chalk, A. F., 204
Change, mutability, motion, flux, 71–75, 79, 82, 85–86, 97–100, 102–103, 114, 116–119, 122, 125–127, 132, 141, 142, 143, 145–146, 150, 152, 154, 157, 162–163, 180–181, 185. *See also* Order
  and Eliot, 125–127
  and Filmer, 154
  and Hooker, 98–99
  and Mulcaster, 86
  and Selden, 154
  and Smith, 73
  and variety, multiplicity, 71–75, 79, 85–86, 98–100, 140, 152
Charles I, king of England, 28, 124, 130
Chivalric values, attitudes, 82–83
Civil War (English), 28–29, 103
Coke, Sir Edward, 35, 198
Colet, John, 72
Collins, Stephen L., 170
Collinson, Patrick, 171
Commerce, trade, 127–138, 179
  free trade, 137
Common good, social good, public good, public weal, theory of, 33, 87, 107, 109, 112, 114, 123–126, 133–134, 156, 158–159, 161, 167, 192
Commonplaces. *See* Order
  in *Gorboduc*, 41–42
Commons. *See* Parliament
Commonwealth, idea of, 18, 23, 32–33, 35–36, 38, 73–75, 78–79, 88–89, 94–96, 104–105, 107–108, 110–114, 117, 119, 123–125, 129, 130–134, 154, 156, 161–162, 164, 172, 192
  and merchants, 128, 130–134
  and Smith, 73
  and Wentworth, Peter, 88
Compact. *See* Contract theory
Complementarity, 13, 98, 117, 164, 187
Consciousness, 4, 5, 24, 41, 57, 72, 83, 86, 101, 103, 165–166
  history of, 5, 165
  and Marlowe, 42
  and order, 4, 5, 24, 165–166
  of other, 101
  of self, 5, 9, 52, 61, 67, 70, 78, 86, 101, 109, 119
  self-consciousness of, 5, 60, 165, 168
  and Shakespeare, 41, 52
  and social reality, 4, 24, 165–166
Constitution, 88–89, 125, 151–152, 155, 157–159, 161–163
Constitutional monarchy, 155, 158
Contract theory, social (covenant, compact), 34, 91–92, 94, 96–97, 105, 155–157, 161–164, 192
  and Hooker, 94, 96–97
Contracts, 107, 119, 136, 151, 153–154. *See also* Contract theory
  and Selden, 151, 153–154
Coronation oath, 15, 105, 107–108
Correspondence theory. *See* Order
Covenant. *See* Contract theory
Cowell, John, 115, 199–200
  The Interpreter, 115, 199–200
Cragg, Gerald, 194
Craig, Sir Thomas, 77, 103, 106–108
Cranfield, Lionel, 129–130
Cranmer, Thomas, 20
Crompton, Richard, 16, 20, 76, 82–83
Cromwell, Thomas, 20, 158
Culpeper, Thomas, 136
Cultural history, 165
Cycles. *See* Time

Decorum, 24, 179, 187
  and identity, 24, 179
Dee, John, 10, 69, 87, 100, 113, 126, 139, 140, 207
Dewar, Mary, 79, 203
Digges, D., 134
Digges, Duddley, 155–156, 211
Discourse, 5, 101, 150, 164, 166, 187–188
  history of, 5
Divine cosmos, 6, 16, 39, 41, 111, 118–119, 145, 168. *See also* Order
Divine right theory, 75, 90, 102, 105–108, 155
  and Doleman controversy, 105–108
  and James I, 110
Doleman. *See* Parsons, Robert
Doleman controversy, 103–108, 197
  and Catholic political thought, 103–106
  and divine right theory, 105–108
  and law, 104–108
Dollimore, Jonathan, 170–171, 175, 184, 186, 205
Donawerth, Jane, 184
Donne, John, 4, 141
  "The Anniversaries," 141
  Sermons, 141

Driscoll, James P., 185, 189
Duty. *See* Order

Eagleton, Terry, 61, 184, 186–187, 205
East India Company, 134
Economic crisis, 127–128
Economy, economic activity, 127–131, 134–135, 137–138
Education, 86–88, 140, 179, 191
Edward II, king of England, 108
Edward III, king of England, 77
Eirenicist vision, 11, 140
Election theory, 103–104, 106, 167
Eliot, Sir John, 123–127, 138, 139, 201–202, 206
 De Jure Majestatis, 123–125
 The Monarchy Of Man, 124–127
 and sovereignty, 123–125
 and time and change, 125–126
Elizabeth, queen of Bohemia, 11, 142
Elizabeth, queen of England, 19, 28, 75, 80–82, 88–90, 111, 115, 122, 124, 129, 142, 169, 179, 207
 and identity, private/public, 80–82
 and succession, 80–81, 88–90
Elizabethan World Picture, 8, 11–12, 15, 96–97, 169, 184
Elton, G. R., 174
Elyot, Thomas, 16–18, 21–27, 73, 75, 87, 191
 The Book named The Governor, 18
Eschatological order, 6, 143–145, 149, 167, 172. *See also* Hobbes, Thomas
Essential order, meaning, 31, 97–98, 108, 144–145, 149–150, 152, 155, 162, 207
Essex, Robert Devereux, second Earl of, 27, 82–83, 141
Exclusion crisis, 103
Existential (reality, truth, order, meaning), 13, 31, 55, 60, 80, 97–99, 145, 151, 153, 161–162, 166, 170, 187, 207
 and Bacon, 80
 in *Coriolanus*, 66, 68
 in *Macbeth*, 64–65
 in *Troilus and Cressida*, 60

Fables, 119, 146–147
Fallen nature, The Fall, 76–77, 84, 143
Faulkner, Robert K., 193
Febvre, Lucien, 186
Felicity, 37, 39, 83–85, 117, 125–126
Fenton, Roger, 136
Ferguson, Arthur, 93, 97, 102, 180–182, 188, 189, 193, 195, 200, 210
Ferne, Henry, 155–157, 159, 211–212
Ferry, Anne, 185
Ficino, Marsilio, 126, 139, 140, 195
Figgis, John Neville, 110, 113, 197
Filmer, Robert, 13, 109–110, 113, 136–137, 149–155, 157, 209–211
 "The Anarchy of A Limited or Mixed Monarchy . . . ," 149
 and change, 154
 and law, 154
 and order, 150, 153
 Patriarcha, 136, 150, 209
 and patriarchy, 150, 152–154
 and property, 150, 152
 as radical theorist, 150–153
 and secular sovereignty, 154
 and state, 150, 152
 and usury, 136–137
 and will, 151–153
Florentine academy, 72
Floyd, Thomas, 77–79, 82
Flux. *See* Change
Forset, Edward, 109–110, 114, 139, 196, 199, 206
 A Comparative Discourse of the Bodies Natural and Politique, 113
Fortune, 118, 120, 122, 200
Francesco, Giorgio, 11
Frederick V, Elector Palatine, King of Bohemia, 11
Free speech, 88
Free will, 77, 84, 112
French, Peter, 139, 140, 206
Freud, Sigmund, 154, 168
Fromm, Eric, 190
Frye, Northrop, 55, 60, 187, 207
Fulbeck, William, 74, 77, 79, 119, 134
Fulbrook, Mary, 171

Gardiner, S. R., 199
Gaveston, Piers, Earl of Cornwall, 108
Gay, Peter, 170
Gibson, Anthony, 207
Goldberg, Jonathan, 198–199, 210
Golden age, 140–143
Good counsel. *See* Order
Goodness, idea of, 72, 76, 100–102, 146, 195
Government obligation. *See* Obligation, government, political
Gray, Charles, 198
Great Chain of Being, 11, 16, 97
Greenblatt, Stephen, 173–175, 178–179, 184, 186
Greenleaf, W. H., 91, 110, 172, 197
Greville, Fulke, 140, 141, 142, 177
 Life of Sidney, 140–141, 142
Grislis, Egil, 193
Grotius, Hugo, 156

Guicciardini, Francesco, 200
Gunn, J. A. W., 173, 203

Harvey, William, 36
Hathaway, Michael, 207
Hayward, Sir John, 103, 106–108, 139, 197
Heidegger, Martin, 56
Heller, Erich, 47, 60
Henry VIII, king of England, 15, 178
Henry (Stuart), prince of Wales, 111–112
Herle, Charles, 155, 157, 159, 160–162, 211, 213
Hermeticism, 11–12, 70, 113, 119, 138–143, 145, 147, 174–175, 195–196, 199, 206–208
  and order, 11–12
  and Renaissance magic, 11, 139–142, 145
Hesiod, 140, 213
Hill, Christopher, 196
Hill, W. Speed, 193
Hilton, Walter E., Jr., 175, 191–192
Historical psychology, 5, 8, 26, 166, 173
Hobbes, Thomas, 4, 6, 13, 25, 28–39, 66–67, 69–70, 73–76, 78–79, 84–85, 91, 94, 100–101, 108, 117, 120–125, 129, 145, 146, 149–154, 156–158, 163–164, 167–168, 172, 175–176, 178, 180–181, 183, 187, 192–194, 200, 202, 213
  and appetites and aversions, 121, 164
  and Bacon, 120, 122, 145, 146
  and the commonwealth, 32
  De Cive, 32, 36, 38
  Elements of Law, 34–35, 156
  and eschatology, 6, 37, 149, 167, 172
  and evaluation, values, 37–38
  and existential truth, 13
  and Hooker, 98, 100
  and idea of covenant, 34
  and idea of order, 5–6, 29, 149, 164, 167
  Leviathan, 28–29, 31, 34–35, 37, 149, 163, 167
  and merchants, 129
  and natural law, 35, 163
  and reason, 35–37
  and representation, 34
  and the secular sovereign state, 13, 28–30, 34, 149–150, 164, 167
  and self-interest, 79
  and Shakespeare, 164
  and theory of natural liberty, 29, 163–164
  and Tudor commonplaces, 37–38
  and will, 164
Holmes, Peter, 197
Homer, 140
Homilies, 15–16, 18–19, 26, 87, 191
Hooker, Richard, 20, 40, 72, 91–103, 108, 125, 139, 159, 161, 169, 178, 193–194, 206
  and authority, 95–98
  and authority/order relationship, 95–97
  and contract theory, 94, 96–97
  and Hobbes, 98, 100
  and law, 20, 92–94, 96
  and Marlowe, 99–100
  and More, 100
  Of the Laws of Ecclesiastical Polity, 91–103
  and order, 94–97, 101
  and relationship of reason/will, 102
  and secular order, 101
  and Shakespeare, 98, 100–101
  and theory of motion, 98–99
  and Thomism, 98–99
Hoopes, Robert, 180
Huizinga, Johan, 23, 173, 190
Humanism, 16, 72, 87, 170, 189
Hunton, Philip, 157–159, 161, 211–212
  A Treatise of Monarchie, 157

Idealism, 170
Identity, 5, 6, 10, 21, 24, 29, 71, 75, 78, 80–82, 85–86, 112, 115, 119, 121–123, 132, 133, 135, 138, 147, 162, 164, 166, 171–173, 179, 181, 185, 187, 189, 198
  and Elizabeth, 80–82
  in specific works:
    As You Like It, 50–51
    Coriolanus, 65–66, 68
    Edward II, 48
    Gorboduc, 42
    Hamlet, 61–62
    Henry IV, 54–55
    Julius Caesar, 56–58
    King Lear, 63
    Macbeth, 64–65
    Richard II, 52–53
    Tamburlaine the Great, 43
    Troilus and Cressida, 58–60
  and tragedy, 55–56
Impetus theory, 99
Intellectual history, vii, 4, 6–8, 29, 92, 110, 165–166
  of order, vii, 4
  of political thought, 110

Jacobean court, 142
James VI and I, king of Scotland and England, 12, 28, 77, 83, 103, 106, 108–116, 124, 129, 142, 156–157, 198–200
  "Basilikon Doron," 111
  and divine right theory, 110
  and law/will relationship, 112
  and merchants, 129
  and the state, 111

Jaspers, Karl, 49, 186
Jaynes, Julian, 213
Jewel, John, 193
Joachim of Fiore, 143
Jones, Inigo, 142
Jones, W. R. D., 28, 38, 172
Jonson, Ben, 142
Jordan, W. K., 127, 129, 182, 202
Judson, Margaret A., 157, 161, 182, 210, 212–213
Justices of the peace, 77

Kaufmann, R. J., 187
Kelso, Ruth, 207
Kermode, Frank, 174
Kleist, Heinrich von, 143
Knowledge, concept of, 138, 139, 140, 143–145, 146, 147–148
Knox, John, 176
Kramnick, Isaac, 172
Kuhn, T. S., 7–9, 165–166, 172–173, 189–190

LaCapra, Dominick, vii
Lambard, William, 77
  Eirenarcha, 77
Lane, R. Todd, 202
Language, 165–166, 188, 198, 213
Laslett, Peter, 150
Law, 20, 32, 35–36, 73–77, 92–94, 96, 99, 104–108, 109, 111–112, 114, 115, 117, 123–125, 149, 151–154, 159–163
  canon law, 93
  ceremonial law, 76, 93
  civil law, 20, 25, 35, 38, 76, 105, 124, 136, 150, 178
  common law, 94, 107, 154, 157, 178
  divine law, 76, 105, 134, 151–152
  and Doleman controversy, 104–108
  and Filmer, 154
  and Hobbes, 163
  and Hooker, 20, 92–94, 96
  human law, 93, 105, 163
  and James I, 112
  judicial law, 76, 93
  law of nations, 74
  law of nature, natural law, 20, 23, 27, 35, 38, 71, 73, 79, 92, 104, 106–108, 117, 134, 148, 163, 183
  to "make" law, 77, 93, 154
  moral law, 76
  and order, 20, 76–77, 93, 105, 123–124
  positive law, 92–93, 96, 104, 108
  relativity of, 93
  scriptural law, 93
  secular law, 76–77, 96
  and Selden, 151, 153
  sovereignty of, 77, 92, 108, 112, 124–125, 158, 167
  and the state, 92
  statute law, 20, 27, 157, 178
Levao, Ronald, 41, 175, 206
Levin, Harry, 49
Liberty, 33, 77–78, 88–89, 111, 116, 122, 151–152, 154, 158–160, 163–164
  in Hobbes, 163–164
  natural, 77, 153
  in Parker, 160
  parliamentary, 116
Limited monarchy. *See* Constitutional monarchy
Locke, John, 92, 150, 210
Lovejoy, Arthur, 9, 45, 146, 189, 195
Lund, William R., 182
Lupercalia, 187

McAlindon, T., 179, 187
Machiavelli, Niccolo (Machiavellian), 73, 97, 110–111, 114, 117, 135, 199–200
McIlwain, C. H., 197
McNeilly, F. S., 213
Macpherson, C. B., 184
Magic, 45, 81–82, 139–142, 146–147
  actives and passives, 81–82, 101, 114, 141, 142. *See also* Hermeticism; Shakespeare
  in *Doctor Faustus*, 45
  in *Julius Caesar*, 57–58
  in *The Tempest*, 69–70
Magician. *See* Magus
Magus, 10–12, 72, 138, 139, 145, 168
Malynes, Gerard de, 128–134, 175
  The Antient Law-Merchant, 132
Manley, Frank, 206
Mariana, Juan de, 155
Marlowe, Christopher, 3, 8, 12, 26, 37, 40–49, 67–68, 72, 79, 82–85, 99–100, 139, 145, 146, 175, 186, 200, 208
  and consciousness, 42
  and Hooker, 99–100
  and meaning, 49, 67
  and order, 41, 49
  and salvation (redemption), 4, 42, 49
  and secular values, 42
  and self-consciousness, 67
  and self-definition, 42
  and values, 49

Marlowe, Christopher (*continued*)
  works:
    Doctor Faustus, 3, 44–47
      and Christianity, 44–45
      and magic, 45
      and mortality, 44, 46
      and order, 44, 47
      and salvation, 44–47
      and secular values, 44
      and self-consciousness, 45–46
    Edward II, 27, 47–49
      and relation between private and public identity, 48
      and self, 48
    The Jew of Malta, 47–48
    Tamburlaine the Great, 40, 42–45
      and identity, 43
      and order, 43
      and self-definition, 42
Marshall, John S., 193
Marten, Sir Henry, 117
Marx, Karl, 154
Mary, queen of England, 17
Mary, queen of Scotland, 88
Masinton, Charles S., 185
Mason, Robert, 24, 135, 136, 207
Masques, 142, 208
Mayer, Thomas F., 176
Meaning, vii, 8–9, 12, 21, 38–40, 53, 56, 82–83, 85, 92, 98–99, 108, 115, 121–123, 127, 138, 142, 143, 146, 151, 159, 167, 170
  in *Coriolanus*, 65–66, 68
  as cultural production, 8, 12, 40, 60
  in *Hamlet*, 61–62
  and Marlowe, 49, 67
  and order, vii, 9, 14, 24, 56
  redefinition of, viii, 9, 12–13, 28, 61, 67, 71, 75, 78, 85, 101–102, 110, 125, 130, 132, 139, 141–142, 149, 170
  in *Richard II*, 53
  and self, identity, 24, 75
  and Shakespeare, 41, 67
  and the sovereign state, 28, 39
  and the state, 83
  and tragedy, 56
  in Troilus and Cressida, 58, 60
Merbury, Charles, 75
Merchants, 12, 27, 111, 127–138, 175, 179
  and the commonwealth, 128, 130–134
  and Hobbes, 129
  and James I, 129
  pamphlets and literature, 127–137, 203
Merleau-Ponty, Maurice, 165
  "Materials for a Theory of History," 165
Mirror for Magistrates, 22, 26
Misselden, Edward, 129–131, 132, 133, 175
  Circle of Commerce, 129–130
  Free Trade, 129, 130
Mixed government, idea of, 15, 27, 71, 155, 161
Mixed monarchy, 152, 162–163
Modernity, viii, 4, 5, 165, 168, 171
Montaigne, Michel de, 120, 183
Montrose, Louis Adrian, 184–186
More, Sir Thomas, 72, 100, 209
  and Hooker, 100
  Utopia, 100, 209
Moretti, Franco, 169, 176, 194
Morison, Richard, 17, 19
  A Remedy For Sedition, 17
Morris, Christopher, 177
Mortality, 3–4, 39, 142, 146
  in *Doctor Faustus*, 44, 46
Moses, 93, 137
Mosse, George, 189
Mosse, Miles, 135–136
Motion. *See* Change
Muchmore, Lynn, 203
Mulcaster, Richard, 86–87, 90–91, 191
  as Baconian, 87
  and change, 86
  and education, 86–88, 191
  and time, 86
Multiplicity, 72, 100. *See also* Change
Mun, Thomas, 132, 133, 134, 136, 137–138, 175
Munz, Peter, 193
Mutability. *See* Change; Shakespeare

Nature, concept of, 139, 140, 141, 143–148, 152–153, 156, 163–164
Neale, Sir John, 88, 90
Necromancy, 141. *See also* Magic
Nef, John, 128
Neoplatonism, 11, 70, 100, 113–114, 118–119, 141, 143, 147, 199. *See also* Hermeticism
Nominalism, nominalist attitudes, 73–74, 101, 151
  and Smith, Sir Thomas, 73–74
Norbrook, David, 178
Norton, Thomas, 3–4, 41. *See also* Sackville, Thomas

Oakeshott, Michael, 36, 167
Oaths, 107, 119, 151, 162. *See also* Coronation oath
Obedience, 93–95, 151
Obligation, government, political, 150–153
Obligation, theory of, 109, 157–158, 163–164

Oliphant, James, 191
Order (idea of; ideology of), vii, 4–11, 13–17, 20, 24–29, 38–39, 53, 55–56, 61, 71–78, 80, 82–83, 85, 87–90, 92–94, 101–107, 109, 112–119, 121, 123, 126, 127, 128, 134–135, 137–139, 140, 142, 144–146, 149–153, 155–156, 158–169, 170, 184–185, 188, 192, 194
  and consciousness, 4, 5, 24, 165–166
  as dynamic concept, viii, 15–16
  and Filmer, 150, 153
  and history, 166, 168
  Hobbes', 5, 6, 29, 149, 164, 167
  and Hooker, 94–97, 101
  and identity, 41
  and Marlowe, 41, 49
  and meaning redefinition, 9
  and paradigms, 7, 13
  and Parker, 30–34, 162
  and representation, 34, 165–168
  and Shakespeare, 40–41, 70
  and social reality, 4, 24, 165–166
  and sovereign state, 7, 83, 109, 123, 149, 162
  in specific works:
    As You Like It, 50–51
    Coriolanus, 66–67
    Doctor Faustus, 44, 47
    Henry IV, 54
    Julius Caesar, 57–58
    King Lear, 63
    Richard II, 52–53
    Tamburlaine the Great, 43
    The Tempest, 69–70
    Troilus and Cressida, 14–15, 58
  and tragedy, 55–56
  Tudor (traditional), viii, 3, 5, 6, 14–16, 22, 24–29, 32, 39, 87, 91, 95–97, 100, 102–103, 105–106, 110, 112–113, 116, 119, 121, 123, 131, 138, 140, 145–146, 150, 152, 166, 170, 176
    and ceremony, 20
    and commonplaces, 3, 8, 14, 17–19, 23–24, 36, 37, 42, 71, 73, 78, 86, 96–97, 99, 101, 111, 119, 123, 132, 148, 159
    and correspondence theory, 16, 20, 86, 89, 97–99, 110, 116–117, 123–126, 139, 145, 148, 151, 155, 159, 162
    and cult of authority, 15
    and divine cosmos, 16
    and duty, 19, 22, 153
    and good counsel, 25, 42, 75, 79
    and great chain of being, 16
    and law, 20, 76–77, 93, 105
    and mutability, 26, 42, 73, 99–100
    and self, identity, 20–21
    and Wentworth, Peter, 88–90
Orgel, Stephen, 207
Origins of government, society, 94, 112, 149–150, 152
Orpheus (orphic magic), 140. *See also* Hermeticism; Magic

Palmer, D. J., 186
Paracelsus, 145
Paradigm, vii, 7–8, 13, 27, 90, 165–166, 169, 172–174, 176, 189, 213
  and meaning redefinition, 7
Parker, Henry, 13, 30–34, 38–39, 155, 159–162, 182, 211–213
  and the common good, 33
  Jus Populi, 31
  and liberty, 33, 160
  and order, 30–34, 162
  and parliament, 34, 161–162
  and power, 30–31, 159, 161
  and prerogative, 33, 160
  and representation, 34
  and secular sovereignty, 31, 161–162
  and sovereignty, 34, 161–162
Parliament, 12, 18, 20, 26–27, 31, 33–34, 72, 81, 88–91, 95–97, 110–112, 115–116, 125, 150, 152–153, 155, 157–163, 174, 177, 183, 192
  as the body politic, 18
  Commons, 88–90, 115–116, 152–153, 162, 175
  as court, 18, 20
  King-in-Parliament, 88, 162
  and monarch, 88
  and representation, 18, 34, 88
  sovereignty of, 26, 102–103
  of 1559, 81
  of 1571, 88
  of 1576, 88
  of 1603, 111
  of 1609, 110
Parsons, Robert (Doleman, R.), 94, 103–108, 197
  A Conference about the Next Succession to the Crown of England, 103–108, 151
Paternity. *See* Patriarchy
Patriarchy (paternity), 108, 137, 150–156, 164, 197, 209
  and Filmer, 150, 152–154, 209
Perception, history of modes of, 165
Pico della Mirandola, 11, 113, 119, 126, 139
Pimander, 207
Pitkin, Hanna Fenichel, 177, 183
Plato, 86, 125
Platt, Gerald, 214

Plenitude, 72
Pocock, J. G. A., 6, 169, 171–174, 176, 188, 198, 213
Political obligation. *See* Obligation, governmental, political
Ponet, John, 17, 20–21, 76, 176
Popery, papistry, 110, 155. *See also* Catholic; Doleman controversy
Porter, H. C., 194
Power, acquisition of, 149
Prerogative (royal), 12, 16, 33, 88, 90, 95, 115–116, 124, 152, 160
Presbyterianism, 110, 154. *See also* Puritanism
Prime mover, 99
Private/public relationships, 21, 75, 79, 80–83, 85–86, 94–95, 109, 112, 122–123, 129, 130–132, 133, 156, 164, 173, 198. *See also* Shakespeare, (works)
Property, 150, 152, 154
  and Filmer, 150, 152
  and Selden, 150
Prynne, William, 152
Public good. *See* Common good
Public weal. *See* Common good
Puritanism, 5–6, 12–13, 80, 86, 90–92, 104, 171, 174, 181, 192–193
Pym, John, 149
Pythagoras, 139

Rabbards, Ralph, 207
Rabkin, Norman, 184, 185, 187
Raleigh, Sir Walter, 4, 33, 35, 116–117, 119, 122, 123, 126, 142, 146, 199
  and law, 35
  Maxims of State, 33
  and time, cycles, 117
Ratcliffe, Ægremont, 19, 27, 79
Reason, 24–25, 35–37, 90, 95, 98, 102, 106, 121, 144, 163
  right reason, 24, 36–37, 94, 121, 144, 194
Rebholz, Ronald A., 178
Rees, Joan, 178
Representation, 6, 18, 34, 105, 150, 153, 165–168, 177, 183
  legislative, 105
  and order, 150, 165–168
  and parliament, 18, 34
  and substitution, 18, 150, 153, 167
Ribner, Irving, 185
Richard II, king of England, 15
Ripley, G., 207
Roberts, John, 137
Roberts, Lewes, 133, 134, 137, 138
  Treasure of Traffike, 137
Robinson, Henry, 132, 134, 138
Ronan, Clifford, 187
Rosicrucian vision, order, 138–140, 142, 147, 196, 199, 206. *See also* Hermeticism; Magic
Rossi, Paolo, 142, 207
Rous, Francis, 30
  The Lawfulness of Obeying the Present Government, 30
Royal Society, 138
Royalists (thought), 150, 152, 155, 157–158, 172, 211

Sackville, Thomas, 3, 4, 19, 22, 26, 41, 73
  "Induction to Mirror for Magistrates," 26
  with Thomas Norton, *Gorboduc*, 3, 18–19, 41–42
    and identity, 42
    and Tudor commonplaces, 41–42
Savage, Alice L., 170
Schiller, Friedrich, 143
Schochet, Gordon, J., 197, 209–210
Science, 138–139, 206
Secular (values, attitudes), 5–6, 30, 37, 39, 72–74, 76–77, 81–83, 89, 101, 110–112, 114, 117, 119–120, 121–122, 127, 128, 129, 134–138, 143, 148, 154, 159, 168
  in *Doctor Faustus*, 44
  in Hooker, 101
  in Marlowe, 42
  in Smith, 73–74
  in Wentworth, Peter, 89
Selden, John, 13, 150–151, 153–154, 156, 176, 210
  and change, 154
  and contracts, 151, 153–154
  and law, 151, 153
  and property, 150
  and secular sovereignty, 154
Self, 5, 6, 8–10, 21, 24, 78, 81, 84–85, 87, 100, 110, 120–123, 126–127, 138–139, 141, 154, 164, 167–168, 173–174, 179, 186–187, 198
  and the divine order, 21
  as gnostic force, 5, 167, 170
  history of, 8
  and identity, 21, 24
  and Marlowe, 42, 45–48
  relation with other, society, 64, 83, 91, 121–123, 138–139, 162–164, 166–167
  and salvation, 5

self-interest, 9, 79–80, 82–83, 85, 87, 95, 134, 155–156, 160, 167
  in specific works:
    Coriolanus, 65, 67–68
    Edward II, 48
    Hamlet, 61–62
    Julius Caesar, 56
    King Lear, 62–63
    The Tempest, 68, 70
    Troilus and Cressida, 58–60
Self-consciousness, viii, 9, 27–28, 40, 45, 78, 83, 85, 89, 102, 111, 115, 117, 118, 127, 129, 140, 161, 166, 168, 175, 209
  and Marlowe, 67
  and Shakespeare, 54, 56, 67
  in specific works:
    As You Like It, 50–51
    Coriolanus, 65
    Doctor Faustus, 45–46
    Henry IV, 54
    Julius Caesar, 57
    King Lear, 63
    Macbeth, 64
    The Tempest, 69–70
    Troilus and Cressida, 59–60
Self-definition, 42, 60, 79–81, 84, 86, 100, 109–112, 122–123, 152, 159, 166–167, 186
  and Marlowe, 42
  and Shakespeare, 70
  in specific works:
    Coriolanus, 67–68
    Julius Caesar, 57–58
    Macbeth, 64–65
    Tamburlaine the Great, 42
    The Tempest, 70
Self-fashioning, 174, 186
Shakespeare, William, vii, 4, 8–10, 12–14, 20, 22, 37, 40–41, 49–70, 72, 79–80, 82–84, 91, 97–101, 105, 117, 119, 120, 121, 122, 123, 141, 142–143, 146, 164, 166, 168, 175–176, 181, 184, 186–187, 189, 191, 198–199, 202, 205–206, 209, 213
  and Bacon, 121, 122–123
  and complementarity, 13, 98, 117
  and golden age theme, 141–143
  and Hobbes, 164
  and Hooker, 98, 100–101
  and identity, 41
  and meaning redefinition, 41, 67
  and order, 40–41, 70
  and self-consciousness, 54, 56, 67, 70
  and self-definition, 70
  works:
    As You Like It, 9–10, 49–51
      and identity, 50–51
      and order, 50–51
      and self-consciousness, 50–51
      and time, 51
    Coriolanus, 10, 62, 65–68, 105
      and consciousness of self, 67
      and existential reality, 66, 68
      and identity, private/public, 65–66, 68
      and meaning, 65–66, 68
      and order, 66–67
      and sacrifice, 67–68
      and self, 65, 67
      and self-consciousness, 65
      and self-definition, 67–68
      and self/other relationship, 65
      as Hobbesian, 66–67
    Hamlet, 10, 27, 55, 60–62
      and identity, private/public, 61–62
      and meaning, 61–62
      and self, 61–62
    Henry IV Parts I and II, 9, 51, 53–55
      and identity, 54–55
      and kingship, 55
      and order, 54
      and ritual, 54–55
      and self-consciousness, 54, 56
      and time, 54
    Henry VIII, 142–143
    Julius Caesar, 10, 52, 55–58, 61
      and identity, private/public, 56–58
      and magic (ritual), 57–58
      and order, 57–58
      and sacrifice, 57–58
      and self, 56
      and self-consciousness, 57
      and self-definition, 57–58
      and values, 58–59
    King Lear, 10, 19, 62–63, 120
      and identity, 63
      and nature, 63
      and order, 63
      and sacrifice, 63
      and self, 62–63
      and self-consciousness, 63
      and values, 63
    Macbeth, 10, 40, 62–65
      and existential reality, 64–65
      and identity, 64–65
      and self-consciousness, 64
      and self-definition, 64
      and self/other relationship, 64
    A Midsummer Night's Dream, 9–10, 49–50

Shakespeare, William,
works: (*continued*)
Richard II, 9, 51–54
and consciousness of self, 52
and identity (relation between private and public self), 52–53
and kingship, 52
and meaning, 53
and order, 52–53
The Tempest, 10–11, 40, 68–70, 141, 175, 188–189, 199
and comparison with Hobbes, 69–70
and magic, actives/passives, 69–70, 141
and nature, 69–70
and order, 69–70
and reciprocal relations, private/public, 68–70
and redemption (salvation), 69
and self-consciousness, 69–70
and self-definition, 70
and self/other relationships, 68, 70
Troilus and Cressida, vii, 10, 14–15, 37, 41, 55, 58–61, 67, 84, 97, 121, 187, 205
and existential reality, 60
and idea of order, 14–15, 58
and identity, 58–60
and meaning, 58, 60
and self, 58–60
and self-consciousness, 59–60
and time, 59
and values, 15, 58–59
The Winter's Tale, 68
Ship money, 159–160
Shirley, F. J., 94, 102, 193–194
Sidney, Sir Philip, 11, 80, 139, 140, 147, 177, 206
The Defence of Posey, 140
Skinner, Quentin, 181–182
Smith, Lacey Baldwin, 172–173, 178–179, 191
Smith, Sir Thomas, 18, 72–75, 79, 94, 103, 128, 131, 134, 177, 189, 203
and change, mutability, 73
and commonwealth, 73
De Republica Anglorum, 72, 75
A Discourse of the Commonweal of This Realm of England, 79, 128, 134
and nominalism, 73–74
and self-interest, 79
and temporal, secular values, 73–74
Smith, Thomas, governor of the East India Company, 134
Social contract. *See* Contract theory
Social good. *See* Common good
Social reform, 140, 142, 143, 146, 147, 148
Social utility (public), 6, 12, 86, 88, 110, 114, 118, 125, 127, 139, 143, 148, 156, 160, 164
Sommerville, J. P., 197, 201
Sovereign state, 4, 6–7, 13, 69, 81, 101–102, 149–150, 164, 167–168
and Hobbes, 13, 28–30, 149–150, 164, 167
and order, 7, 83, 109, 149, 162
and Parker, 161–162
secular, 4, 6–7, 13, 30–31, 35, 39, 75–76, 109, 112, 123, 149–150, 155, 158–159, 162, 167, 171, 181, 192
Sovereignty, 16, 26–27, 30–31, 34–35, 90–92, 95–96, 102, 108–110, 114–116, 123, 138, 149–152, 155, 161–162, 164, 167, 178, 181–182
and Eliot, 123–125
of king, royal, 110, 115–116, 158
of law, 77, 92, 112, 114–115, 124–125, 158, 167
of Parliament (Commons), 90, 161–162
St. German, Christopher, 20
Stanford, John, 141
Starkey, Thomas, 15, 17–19, 21, 23, 25, 72, 176
Dialogue, 15, 23
Exhortation to Unity and Obedience, 15
State, theory of, 18, 36, 38, 74, 83, 88–89, 111, 150, 152, 156, 161–162, 167, 192
and Filmer, 150, 152
and James I, 111
and meaning, 83
and order, 83
and Wentworth, Peter, 88–89
Stone, Lawrence, 171, 179, 210
Strong, Roy, 178
Subsidy, 115
Succession, 75, 88–89, 103, 106–107, 109, 115, 117

Talbert, E. W., 177
Tawney, R. H., 128
Tew circle, 211
Thomas, Keith, 175
Thomism (Thomist, Aristotelian), 84, 98–100, 193
Hooker as, 98–99, 193
Thompson, W. D. J. Cargill, 193
Thrupp, Sylvia, 203
Tillyard, E. M. W., vii, 25, 187
Time, 51, 72–75, 82, 86, 111–112, 114, 117–119, 122, 134, 147
in *As You Like It*, 51
and Bacon, 115, 118–119, 120
cycles, 72–73, 78, 117, 125, 126, 142, 147
and Eliot, 125, 126

and Raleigh, 117
temporal (attitudes, values), 37, 39, 72–74, 77–78, 81–83, 99, 106, 120, 145, 160, 166
in *Troilus and Cressida*, 59
Trade. *See* Commerce
Tragedy, 3, 49, 55–56, 61–65, 67–68, 83, 169, 175
and existential reality, 55
and felicity, 83
and identity, 55–56
and meaning, 56
and order, 55–56
and sacrifice, 63
and self, 61–62
Transcendental truth (reality), order, 6, 13, 69, 78, 100, 119, 150, 171
Trismegistus, 113, 195, 207
Tuck, Richard, 176, 181, 183, 209–211
Tuvil, D., 120–121
Essaies Politicke and Morall, 120

Usury, 24, 134, 135–137, 150–151
and Bacon, 136
and Filmer, 136–137

Values, valuation, 9, 15, 23, 37–38, 49, 58–59, 63, 71–72, 74, 79, 83–84, 95–96, 98–99, 109, 121, 127, 129, 130–132, 134, 158, 160
and Hobbes, 37–38
and *Julius Caesar*, 58–59
and *King Lear*, 63
and Marlowe, 49
and *Troilus and Cressida*, 15, 58–59
van den Berg, Kent T., 186
Vergil, Polydore, 106, 108
Veto, 152–153, 159
Virtue, 24, 37, 142, 146
Virtuosi, 175, 191
Voegelin, Eric, 167, 171–172, 174, 183

Waller, Gary, 178
Walsingham, Sir Francis, 19, 27
Walzer, Michael, 5, 29, 91, 192, 169, 181
and Puritanism, 5
The Revolution of the Saints, 5
Warhaft, Sidney, 208
Watkins, J. W. N., 182
Weinstein, Fred, 214
Wentworth, Paul, 88
Wentworth, Peter, 86, 88–91, 94, 192
and commonwealth, 88
comparison with Hobbes, 91
and concept of state, 88–89
and order, 89–91
and Parliament of 1576, 88
and succession, 88–89
Weston, Corrinne C., 212
Wheeler, John, 131–132, 137
Whetstone, George, 17, 22, 78–79
Aurelia, 22
White, Hayden, 170, 172, 195
White, Howard B., 208
Whitehead, Alfred N., 40
Whitgift, John, 193
Williams, Robin M., 173
Wilson, Charles, 203
Wilson, Thomas, 24, 76, 129, 131, 134–135
Discourse Upon Usury, 129, 134–135
Wright, Henry, 117, 119
Wyatt, Thomas, 179

Yates, Francis, 11, 139–140, 142, 147, 170, 174–175, 194, 199

Zeeveld, W. Gordon, 178